MIND AND REALITY

The Space-Time Window

MIND AND REALITY

The Space-Time Window

Wolfram Schommers

University of Texas at Arlington, USA & Karlsruhe Institute of Technology, Germany

NEW JERSEY • LONDON • SINGAPORE • BEIJING • SHANGHAI • HONG KONG • TAIPEI • CHENNAI

Published by

World Scientific Publishing Co. Pte. Ltd.
5 Toh Tuck Link, Singapore 596224
USA office: 27 Warren Street, Suite 401-402, Hackensack, NJ 07601
UK office: 57 Shelton Street, Covent Garden, London WC2H 9HE

British Library Cataloguing-in-Publication Data
A catalogue record for this book is available from the British Library.

MIND AND REALITY
The Space-Time Window

ISBN 978-981-4556-76-7
ISBN 978-981-4556-77-4 (pbk)

In-house Editor: Christopher Teo

Printed in Singapore

PROLOGUE

■ ■ ■

Our observations in everyday life are of basic relevance for the development of scientific conceptions. Within the frame of such direct observations, trees, houses, the moon and a lot of stars appear spontaneously in front of us without any conscious action. Matter seems to be embedded in space and its structure develops in the course of time. But the space is larger than what we experience in everyday life; we have extended this everyday-life view step by step and a huge space-region could be explored systematically. In other words, our world in front of us has been enlarged and we have came to what we call the universe.

We define the universe as the "totality of existence", including planets, suns, galaxies, the elementary particles, dark matter, dark energy, etc. We also extended the time-region in the direction of the past and came to the famous Big Bang theory, which can be considered as the present prevailing cosmological model.

For this kind of universe our everyday-life experiences are of basic relevance, and this is because they reflect the first and most direct interaction a human being has with the world outside. Then, the following question is relevant: What kind of reality do we have in front of us? Is it the objective, basic truth or what kind of reality is it really? Is this everyday-life view in particular independent of the observer's peculiarities? In this monograph we will treat these questions systematically. In order to set the direction for this book, let us here quote a simple experiment, which demonstrates the relevance of these questions.

•

The world before us appears spontaneously without any intellectual help. We consider this "world view" as independent from the observer. This is obviously not the case and is particularly demonstrated by the following experiment: A human being who puts on goggles equipped with inverting lenses sees the entire world upside down, not forever but only initially. After a certain time the entire visual field of the observer flips over and the objects are seen as they had been before the goggles were put on. The process takes place without (conscious) action of the subject.

This simple experiment distinctly reveals that the world we experience spontaneously is not independent of the human observer. The brain ignores the goggles although it belongs to the reality outside. We may in particular conclude that the brain of the subject manipulates the impressions that we have from the outside world, i.e., it is obviously a "constructed world". How can we understand these facts?

What world do we have in front of us? It is obviously this constructed world, designed by the brain on the basis of certain information from the world outside. Already the famous psychologist C.G Jung was impressed by the fact that an event in the world outside produces simultaneously an inner picture. Just this inner picture appears in front of us and is manipulated by the brain when we put on goggles with inverting lenses; the brain constructs unconsciously a world view which is different from the world outside, and this is because the goggles are ignored although they belong to the material world outside.

We know that there are goggles with inverting lenses and we also know that the effect produced by the goggles are ignored by the brain. But what about other entities in the world outside that do not appear before us because the brain ignores them for the construction of the world in front of us? For answering questions like this we have to know how the reality outside is composed. What is its structure and content? For the assessment of the world in front of us, we need in principle the complete information about the reality outside. Otherwise we cannot assess the "level of reality", which is reflected in the picture before our eyes.

Is it the "absolute truth", i.e., the "complete information" about the world outside or is it only a part of it? These questions can be answered when we consider the principles of evolution. Biological systems developed in accordance with these principles, which determined how a human being or another biological system interacts with the world outside. How developed is a human being and his brain during biological evolution? What are the peculiarities of biological evolution?

The perception of the true (absolute) reality in the sense of a precise reproduction implies that with growing fine structure in the world before us (its picture), increasing information of actual reality outside is needed. Then, evolution should have developed the sense organs with the feature to transmit as much information from the outside as possible. But just the opposite is correct: The strategy of nature is to take in as little of the outside world as possible. Reality outside is not assessed by "true" or "untrue" but by "favourable towards life" or "hostile towards life". In other words, it is not "cognition" but the differentiation between "favourable towards survival" or "hostile towards survival" which plays the relevant role.

The brain developed according to these criteria, and this implies that the world in front of us cannot be the absolute truth; it is a brain-dependent "constructed world", as we have already recognized in connection with the goggles with inverting lenses. This restricted information from the actual reality outside is processed by the brain and produces a picture of reality, which appears before our eyes.

This picture is not complete and may contain the information merely in symbolic form; the picture of reality does not have to be complete and true (in the sense of a precise reproduction) but restricted and reliable, at least during the early phylogenetical phase. All these features are controlled by the principles of evolution. So, for example, in order to find a certain place in a cinema, it is not necessary that the visitor gets at the pay desk a small but true model of the cinema, i.e., a one-to-one reproduction of the cinema, which is reduced in size; a simple cinema ticket with the essential information is more appropriate. In this respect, the cinema ticket is the picture of the cinema, like the world in front of us is the picture of the actual reality outside.

•

Newton's world view is essentially based on the world in front of us, i.e., it is based on the direct impressions which we have before our eyes. Cars, houses, the stars and the planets are treated mathematically, resulting in the successfully reliable space-time based description of these and other entities. However, Newton's and other similar world views work on a level at which only restricted information of the world outside is realized. In other words, such theoretical descriptions work on a level at which the picture of reality is not complete and true (in the sense of a precise reproduction) but it is "only" restricted and reliable, and these peculiarities are dictated by the principles of evolution.

Due to these principles only a selected part of the world outside is left available, which is however still reliable. This should also be valid for the mainstream of our theoretical descriptions because they are essentially based on the restricted information selected by evolutionary processes. In other words, the formulas do not reflect the complete world outside but offer reliability. Reliability in particular means that a great variety of situations can be covered by theories. In fact, Newton's theory could be applied successfully to a great variety of situations around us. That is possibly the reason why Newton's theory is deemed "successful"; it has a feature of being at the level of everyday life.

The big success of Newton's mechanics established a world view which became a model for all the other theories, i.e., the basic standards of Newton's procedural method are also reflected in more improved theories like the Theory of Relativity and also partly in the conventional quantum theory. The reason is obvious: The theories are essentially based on observations which the observer experiences in everyday life, which appear directly in front of the observer. This is the basic information for which all further developments stem from. As we have pointed out, the world in front of us can however only be a restricted picture and is particularly species-dependent. In this regard the following is important: *We do not base our theoretical considerations on what really exists in the basic reality, but on what evolution allows us to recognize*. This is Newton's level, and is also the level of the other theories which has been developed subsequently. This species-dependent point of view must in particular be reflected in the statements when we extend the space-time region successively, as we have discussed above. The relatively new notions like "dark matter", "dark energy", "Big Bang theory" and all the other conceptions concerning the basic nature of the universe become therefore uncertain or even useless when we try to recognize absolute standards, i.e., they are only of limited value.

This level corresponds to a world view, which is confined by the reality in front of us (its picture) and how we interpret and assess it, but it is by no means an "ultimate conception".

Alternative conceptions are discussed in this monograph. Let us give here one of the results: At the level of conventional physics, the possibilities for travelling through space are firmly determined. In this case only restricted space regions can be reached within reasonable time intervals, and this is because the velocity cannot exceed the velocity of light. However, other physical conceptions, discussed in this book, could lead to other, less restricted possibilities for travelling through space and time with a spaceship. On the basis of other experimental situations we could possibly reach any place in the universe at any time, that is, we will not only be able to visit "any" space position in the cosmos, but we could also travel in time and could visit "any" point in time in the past, present and future. This is a completely new perspective but it has a real background of supporting evidence and is not purely a speculative statement.

•

Due to the effect of evolution, the world view is dependent on the biological system. The philosopher Immanuel Kant thought in this direction, and was firmly convinced that the impressions in front of a human being in everyday life are essentially influenced by his brain. In fact, modern behavior research supports that.

We have as many world views as there are different species, varying in their biological structure. A lot of scientists believe that we are able to develop a "world equation", and they are obviously firmly convinced that we will have this world equation soon. Here, a world equation means the final mathematical code about the complete world (basic reality). However, the principles of evolution speak another language: Since the world view is dependent on the biological system (human beings, animals, etc.), we should have as many "world equations" as there are different biological systems.

•

The book deals with "mind" and its relationship to "reality"; the material aspect of the world is treated realistically in this monograph. When we look out of a window we see the world outside. Space and time can be considered as a "window" because they reflect certain peculiarities of the world outside. The effect of biological evolution on the theoretical description of physical phenomena will be pointed out in detail. Furthermore, the nature of space and time will be analyzed with respect to what we actually observe. In this monograph we will distinguish between the "container principle" (here the world is embedded in space and time) and the "projection principle" whereupon the world is projected onto space and time.

The author is grateful to all colleagues and students at the University of Texas at Arlington for creative discussions.

Wolfram Schommers
Arlington, Texas, USA
Karlsruhe, Germany

CONTENTS

▪ ▪ ▪

Chapter One

BASIC QUESTIONS

■ ■ ■

In natural science, the observations in everyday life are of particular relevance. We normally base our analysis on those events and things which are spontaneous in front of us; we experience these events and things directly without any interpretation. In such situations we have humans, trees, cars, the sun, the moon, stars and a lot of other things (see Fig. 1) directly in front of us and they appear without any intellectual help [1, 2].

1.1 THE SPACE AS CONTAINER

This kind of world is grasped within so-called "assumption-less observations" in everyday life, and this kind of reality is experienced by each human being in the same manner. A typical example is given in Fig. 1. This world, which we often call "material reality" and which is experienced by assumption-less observations, appears to be embedded in space. On this level reality is considered as a "container" in which the masses are positioned where the container itself is identical with that what we consider as space. Let us call this concept "container principle".

Once again, within such assumption-less observations in everyday life certain macroscopic bodies appear: human beings, trees, cars, houses and other things. But these impressions can be refined by more sophisticated observations, that is, we perform measurements. In this

Fig. 1 The image shows the world in front of us within an assumption-less observation, experienced without any intellectual help. The world (material objects) appears to be embedded in space, and this situation defines the so-called container principle.

way we extend our knowledge about how the world and atoms, molecules, elementary particles, galaxies, suns and planets appear. However, all these things do not change the container principle concept which is reflected through our assumption-less observation in everyday life (Fig. 1). The only difference is that the number of entities, positioned in space, seem to be extended and there are not only the macroscopic bodies (trees, etc.) in the container which we observe in everyday life (Fig. 1); there are additions such as atoms, molecules, etc.

1.1.1 Basic Information

We put all these additional things intellectually into the container and develop on this basis theoretical conceptions that work in most cases excellently. For example, we are able to calculate the movements of planets around the sun very precisely.

All these facts, based on assumption-less and other observations as well as on theoretical conceptions, can be called the "basic information" about reality. But is this "basic information" identical with the "absolute truth"? Can a container filled with material entities be considered as the absolute truth?

Fact is that the human observer is caught in space and time. The observations in everyday life (an example is given in Fig. 1) appear in space, and life develops in the course of time. We construct visual and abstract ideas in space and time in a most natural way.

Fact is also that the impressions that we have directly in front of us — before our eyes — are images of the world outside, which are positioned in the brain, and yet these images appear to be in front of us. In other words, Fig. 1 and all the other direct impressions in front of us are merely images of the world outside. Is there evidence for that? Yes, it is. In Appendix A we bring specific experiments which clearly demonstrate that a spontaneous impression in front of us (Fig. 1) cannot be the actual world outside.

It is essential to mention that the images in front of us come not into existence through the information of the world outside alone, but the eye, the optic nerves and the brain work here together. In other words, the impressions which are in everyday life spontaneous in front of us are dependent on the observer itself and cannot be considered as observer-independent. Thus, the above introduced notion "basic information" is observer-dependent. Nevertheless, this "basic information" could still reflect a certain form of "absolute truth".

Up to now we have nothing said about the structure of the material world outside. A short form of Fig. 1 is given by Fig. 2. We come to the container principle if we replace the crosses, the geometrical symbols, in Fig. 2 by real bodies (masses), i.e. the world outside is defined by a space in which two real masses are embedded leading to Fig. 3.

In other words, we have to distinguish between a "world outside" and an "inside world". Within usual (conventional) physics, in particular within Newton's mechanics, it is assumed as a matter of course

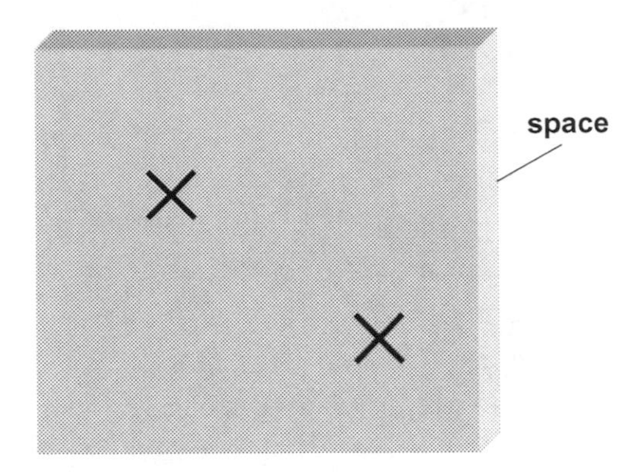

Fig. 2 For the image of reality inside the brain, which appears in front of us, we have chosen a simple form given by two geometrical symbols (two crosses in the figure) that are embedded in space, that is, instead of the complex geometrical structure in Fig. 1 we have two crosses.

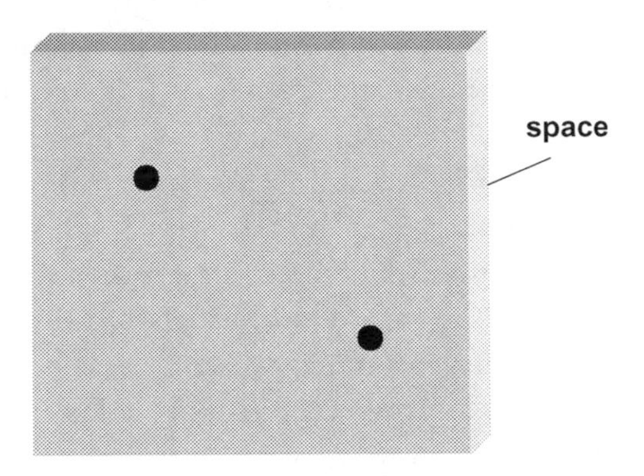

Fig. 3 Within the container principle the world outside consists of a space in which the material bodies (full points in the figure) are embedded.

that there is one-to-one correspondence between reality outside and that what is positioned in our brain (inside world).

We experience the world by our sense organs, i.e., the observer interacts with reality outside: Information about the world outside flow via our senses into the body, and the brain forms an image of the world outside. In other words, we have a reality outside and an inner image of it, and we are usually firmly convinced that the structure of the inner picture (see for example Fig. 2) is identical with that of reality outside (see for example Fig. 3); that is, it is always assumed that there is a one-to-one correspondence. For example, the well-known psychologist C. G. Jung (1875–1961) wrote [3]: *When one thinks about what consciousness really is, one is deeply impressed of the wonderful fact that an event that takes place in the cosmos outside, produces an inner picture, that the event also takes place inside ....*

This statement by C. G. Jung suggests just that what we have already pointed out: The image as well as the real world outside are embedded in space and time.

Clearly, the world in front of us is the inside world but this is not essential because the structure of the inside world (Fig. 2) is identical with that of the world outside (Fig. 3). Therefore, we always assume (consciously or unconsciously) that we have the physically real world outside directly in front of us. We maintain this view in this monograph when we talk about man's usual conception of the world which is given by the container principle. However, we will recognize that this one-to-one correspondence is not realistic and we come to the so-called projection principle.

1.1.2 No Direct Access to the World Outside!

The impression in front of us reflects an inner state within the brain of the observer. The problem is that we have no direct access to the world outside. The simplest assumption is that there is a one-to-one correspondence between the structure in the image (Fig. 2) and the structure of the world outside (Fig. 3), and we come to the container principle. Although this principle is used in conventional physics, the assumption of an one-to-one correspondence seems to be too simple and too naive. Nevertheless, most people believe in this conception, almost always unconsciously; as a typical example we have quoted above the standpoint of C.G. Jung.

Only a few scientists and philosophers treat this problem more critically, but do not come to a definite result. John Borrow remarks [4]: *There are two poles about the relationship between true reality and perceived reality. At one extreme, we find 'realists', who regard the filtering of information about the world by mental categories to be a harmless complication that has no significant effect upon the character of the true reality 'out there'. Even when it makes a big difference, we can often understand enough about the cognitive processes involved to recognize when they are being biased, and make some appropriate correction. At the other extreme, we find 'anti-realists', who would deny us any knowledge of that elusive true reality at all. In between these two extremes, you will find a spectrum of compromise positions extensive enough to fill any philosopher's library: each apportions a different weight to the distortion of true reality by our senses.*

Again, the problem is to find a realistic conception for the world outside, which is not directly accessible to empirical tests. The container principle with its one-to-one correspondence seems to be too naïve, but this statement has to be justified, and we will do that in this monograph step by step.

1.1.3 The Container

The following question arises: Can our immediate experience in everyday life really reflect the "true structure" of the world outside, i.e., the "absolute truth"? Most people would spontaneously say "yes it does" because they firmly assume that the container principle is valid. However, we have to be careful. Fact is that all the things we experience in everyday life (an example is given in Fig. 1) cannot be considered as "absolute truth". In the following we will justify this standpoint, and we will in particular analyze the real situation in detail.

The structure of the world within assumption-less observations reflects a certain level. The levels on which we develop theories and interpretations are in general more sophisticated and also more detailed. However, also at these scientific levels we work within the container principle; in other words, the container principle is not affected when we go from the facts of assumption-less observations to scientific theories. All theories about the world, from Newton's mechanics, to the Theory of Relativity to all forms of conventional quantum theory, have been based on the container principle. But does this principle reflect the real situation? We will recognize that this conception is problematic.

We may state that the container principle is considered in conventional physics as the "basic truth". Whether this basic truth can be identified with the "absolute truth" remains here an open question; it strongly depends on the school of thought. But for most traditional physicists the just defined basic truth is not different from the absolute truth. However, we will recognize below that this standpoint cannot be considered as realistic, and this is because a lot of contradictions appear in connection with the container principle.

To consider the world as a large container (consisting of space and time) in which matter is embedded (Fig. 3) seems to be a prejudice and seems in particular to be too naïve. But this conception is suggested when we interpret our experiences in everyday life, done within the frame of assumption-less observations. Instead we should rather consider space-time as a "window" which enables the observation of the outside reality. We will find out that the real material world should not be embedded in space and space-time, respectively, but should be "projected" on it.

1.1.4 Projections

What does projection mean? Within the container principle the material world is embedded within space. Projection means that the material world is projected onto space (space-time) and we obtain an image of the outside world, i.e., a "picture of reality". Within this conception (let us call it the "projection principle") reality outside does not contain the elements which are space and time.

In other words, within the projection principle the material bodies (real masses) are not embedded in space and time, and this is of course in contrast to the container principle. Clearly, the introduction of the projection principle has to be justified in detail, and we will give the details here.

When we apply such a projection principle consequently we come to a new and surprising relationship between "mind" and "reality". In particular, the role of matter can be analyzed in more detail just by being in connection with the phenomenon of mind. It should be emphasized here that a basic definition of matter within the projection theory can be achieved in a more convincing way than within the framework of the container principle.

Furthermore, one of the reasons for the introduction of the "projection principle" (and for the rejection of the "container principle") is the phenomenon of biological evolution. Evolution obviously prevents a human being from recognizing what is often called "absolute or true reality". In other words, our observed world in everyday life, which appears directly in front of us, cannot be the absolute truth. But what kind of reality can be recognized within the frame of our everyday life experiences? This is an important question that has to be answered on the basis of conclusive arguments.

It is widely accepted that the assumption-less observations in everyday life are the basis for the theoretical analysis of our world. In particular, physical theories may not be in contradiction to that what we have spontaneously in front of us. We have already answered all these question in [1] and [2]; in this monograph we would like to give further arguments and want to deepen this point. In particular the basic features of space and time will be of central relevance when we analyze the situation.

We already mentioned above that the biological evolution is of particular relevance when we analyze our knowledge about the world outside. Concerning biological evolution let us state the following: In traditional physics one tries to understand the phenomenon of evolution with the means of a given physical frame. However, so far we never asked seriously whether the physical frame, i.e., the basic physical laws, are dependent on the laws of biological evolution. We come to a relevant question: To what extent are the physical laws dependent on the laws of evolution? No doubt, this question is relevant and will be answered within in this book.

1.2 TWO SPACE-TIME TYPES?

The main reason for the necessity of the "projection principle" is the phenomenon of biological evolution. This principle can however also be deduced by taking a deeper look into basic space-time features.

Within the container principle, the world outside is embedded in space and time. However, space and time also belong to the observer's mind. When we close our eyes the brain can develop images, i.e., the structures of certain situations, developed by the brain, are represented in space and time.

This in particular means that space and time are tightly linked to the biological system (human being). We know that space and time appear twice within the container principle: In the world inside as well as in the world outside, i.e., we have two types of space-time. This is however problematic and more than questionable. One of the reasons is biological evolution. Because of the appearance of a characteristically distinct space and time within the observer's brain, we have to conclude that the features of the inner space and the inner time are influenced or even constructed by the processes of biological evolution.

Clearly, the features of space and time of the "world outside" cannot be influenced by such factors, i.e., by the biological evolution. Reality outside is assumed to be independent of the observer's peculiarities. In other words, within the container principle two space-time types appear with different peculiarities. This reflects an inconsistency.

Let us mention at this stage that within the projection theory only one space-time type can exist; here only the space-time elements of the inner world are relevant. As we have already remarked above, within the projection theory space and time do not belong to the world outside. This is also in connection with Kant's philosophy, as the famous philosopher Immanuel Kant (1724–1804) came to this conclusion as well. This point of view will be justified and discussed in more detail below.

1.3 INERTIA

1.3.1 The Method of Abstraction

Situations like those that are represented in Fig. 1 are complex and not directly suitable for the formulation of a physical theory. Instead the method of abstraction has to be used for the formulation of physical laws. This method is applied without the general form of appearance (given, for example, in Fig. 1).

In the method of abstraction the material world seems to be embedded in space and space-time, respectively, and this is the view of conventional physics (i.e., within Newton's mechanics, the Theory of Relativity, quantum theory, etc.) and everything is constructed on this basis.

As we know, the methodological action in today's physics is essentially influenced by the conceptions introduced by Galileo Galilei (1564–1642), Johannes Kepler (1571–1639) and Isaac Newton (1643–1727). By mathematical thinking and systematically performed experiments, they established classical mechanics and astronomy, respectively. In particular, with Newton's "Principia" a new interpretation-scheme for natural phenomena had been introduced that is valid up to the present day: Newton investigated the laws of nature by the separation of isolated problems out of the overall context of natural phenomena as they appear, for example, in Fig. 1. As we already remarked, the description of all details given in Fig. 1 is too complex and, therefore, we are only able to analyse specific isolated problems.

We have to underline once again, that the methodological procedures and conceptions in present physics are strongly influenced by the conception introduced by *Isaac Newton*. Newton's mechanics is based on the following conception: We have a space, characterized by the coordinates x, y, z, and there is a time, which we would like to characterize here by the Greek letter τ. Within this space-time N bodies are embedded and each of them has a certain mass. According to Newton, for these N masses there are N equations of motion. As we know, the aim of Newton's equations of motion is to establish laws governing the trajectories of moving bodies.

Let us consider only one body having the mass m and let us assume that this body is a point mass; in this case the position of the mass is given at time τ by the three coordinates $x(\tau)$, $y(\tau)$ and $z(\tau)$. In general, there are forces between the mass m and the other $N - 1$ masses, i.e., interactions described by the gravitational law are effective. This situation holds of course for all the N masses. We already remarked that for these N masses there are N equations of motion, and these equations of motion are able to express the N positions as a function of time τ. This is illustrated with the help of a simple two-dimensional example in Fig. 4 for $N = 3$ point-like bodies.

Here we do not go into the details of Newton's mechanics, but let us underline that Newton's theory was unusually successful and is still used extensively; the theoretical results and statements agree

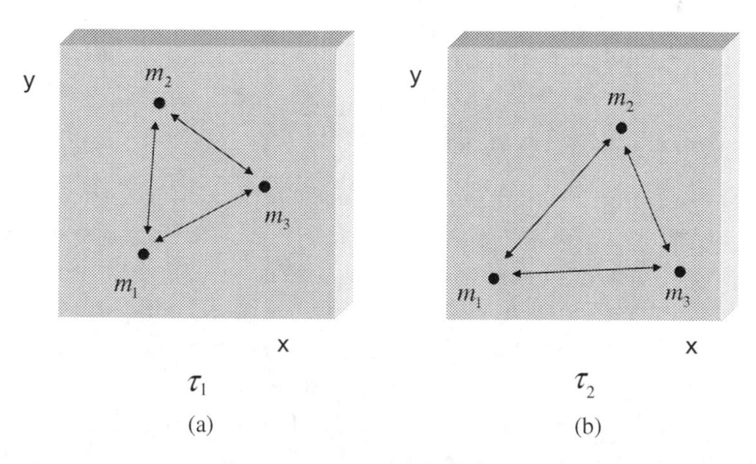

Fig. 4 Three bodies of masses m_1, m_2 and m_3 are embedded within a two-dimensional space having the coordinates x and y. We would like to assume that there is a mutual interaction (as, for example, described by the gravitational law) between the three bodies. Newton's equations of motion allow to calculate the three positions of the three masses as a function of time τ. If the configuration in Fig. 4a is given at time τ_1, we are able to determine on the basis of the equations of motion the configuration at time τ_2 (Fig. 4b). The three bodies move continuously from one configuration to the other.

fantastically with all kind of experimental data. Because of its great success Newton's theory has a reputation of being a scientific revelation. As we already pointed out several times, Newton's mechanics and its principal concepts have become the determining yardstick for the entire field of physics. The essential basic features have even become the peculiarities of the Theory of Relativity and of the conventional Quantum Theory.

From this point of view, the basics of Newton's mechanics — the equations of motion as well as the gravitational law for the interaction between the bodies — suggest that it has to be the (absolute) truth. But we have to be careful: Newton's theory is based on some doubtful elements. We will discuss them in the following, and we will recognize that the "reality" reflected by this theory has to be changed, if these doubtful elements are eliminated.

The notion of "inertia" and the role of "space and time" within Newton's theory have to be considered in more detail. Newton's theory is and was very successful, but its basic elements, in particular the effect of inertia, have to be analyzed critically.

1.3.2 Motion without Interaction

If there are no forces between the masses in Fig. 4 or if there is only one body embedded in space, any mass moves through space with constant velocity: $\mathbf{v} = const$. This effect is due to the inertia produced by the space and this is a feature of Newton's mechanics.

In the case of our example (Fig. 4) we would have $\mathbf{v}_1 = const$, $\mathbf{v}_2 = const$ and $\mathbf{v}_3 = const$, where the three velocities are in general different from each other: $\mathbf{v}_1 \neq \mathbf{v}_2 \neq \mathbf{v}_3$. This situation is represented in Fig. 5a. If we remove body 2 and body 3, only one body (body 1) occupies space; the velocity of body 1 is however not affected and remains $\mathbf{v}_1 = const$ (scc Fig. 5b). If we finally remove also body 1 we get an empty space (Fig. 5c). All these variations are allowed within Newton's mechanics.

What can we say about the situations in Figs. 5b and 5c? What can we say in particular about the reality of space and that of time

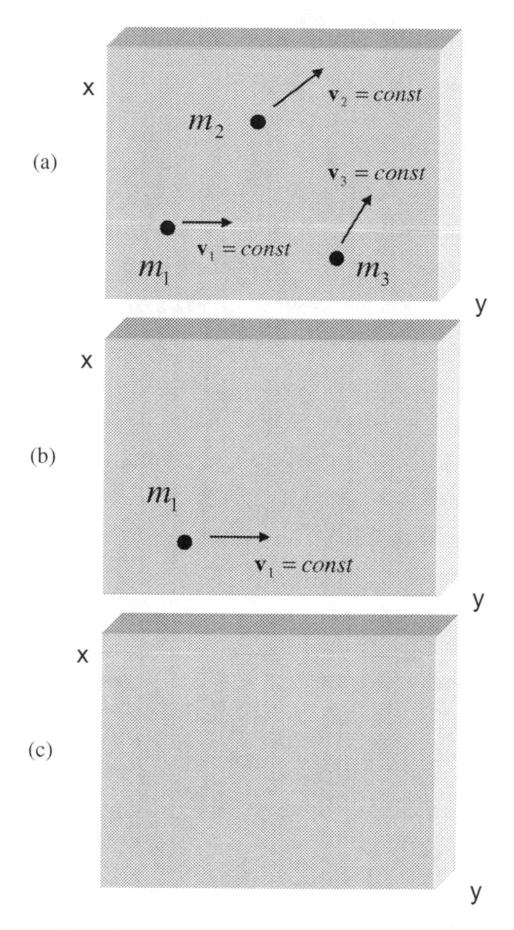

Fig. 5 (a) If there is no interaction between the three bodies we have, in accordance with Newton's mechanics, $\mathbf{v}_1 = const\, \mathbf{v}_2 = const$ and $\mathbf{v}_3 = const$, i.e., the three velocities remain constant in the course of time τ. (b) When two of the three bodies are removed as, for example, body 2 and body 3, only one mass occupies space, but the property $\mathbf{v}_1 = const$ remains conserved. (c) This figure shows an empty space and no material body is embedded in it. Is an empty space a realistic scientific position? This question is answered negatively in the text.

in these cases? Can there exist an "empty space" at all, i.e., a space without real bodies? This must be answered in the negative when we take the point of view of science seriously.

An empty space (Fig. 5b) is principally not observable, and a space with only one body (Fig. 5c) is also not a realistic configuration because it is also not observable. The situations, expressed by Fig. 5b and Fig. 5c, do not reflect scientific realities and are therefore not acceptable. This is important and must be pointed out in more detail.

The behaviour $\mathbf{v}_1 = const$, $\mathbf{v}_2 = const$ and $\mathbf{v}_3 = const$ (Fig. 5a) reflects the phenomenon of inertia, postulated by Newton. Inertia means that the "state of motion" does not change in the course of time if no forces are acting on a mass m. Inertia can only be explained within Newton's mechanics by the space in which the mass m is embedded: There must be a relationship between the space and the mass m, i.e., there is a certain interaction between both elements (space, mass) producing, within the frame of Newton's theory, the effect of inertia. This is however a problematic construction and has to be declined. We would like to point out why this is an ill construction and basically not acceptable to a scientist.

1.3.3 Newton's Space and Time

In Newton's physics, space and time are absolute quantities, and they are independent of each other. They may even exist when space is not filled with matter. In other words, the empty space represented in Fig. 5c is a physically real entity, at least from Newton's point of view. Its existence is necessary because only on the basis of the above mentioned features of space-time Newton was able to construct a reasonable theory of motion. Newton's space is an ad hoc idea and does not withstand any critical scientific analysis. Before we deepen this point, let us explain the significance of this question.

In fact, Newton's space is an artificial construction, and it has nothing to do with a real physical entity; its properties has been postulated but could never be proved. No doubt, the impressions in front of us suggest such a space construction, but this is obviously a fallacy. We have sometimes to live in science with metaphysical elements, but in the case of space and time we need realistic and provable ideas, tailor-made to what we really experience. The reason is simple: The nature of space and that of time determine the basic "world view". In other words, a wrong idea about space and also about time would influence our basic views about the world seriously.

As we have outlined above there must be a relationship between Newton's absolute space and the mass m, i.e., a certain interaction between both entities (space, mass) has to be postulated, producing, in the opinion of Newton, the effect of inertia. This is however a problematic construction and has to be declined. The velocities $\mathbf{v}_1$, $\mathbf{v}_2$ and $\mathbf{v}_3$ defined in Fig. 5a describe the motion of the three bodies relative to this absolute space. Absolute space is an independent container which is filled with material object. But what is space (time) made of? This question is fundamental and needs a clear answer. Let us discuss this point in more detail.

No doubt, Newton's mechanics was a very successful scheme of description, and it is still a useful tool up to the present day. There is presently no alternative to this theory. Einstein (1879–1956) developed the Theory of Relativity, but this theory is essentially based on Newton's basic interpretation scheme. However, an excellent description of important phenomena must not automatically mean that the description scheme (here Newton's theory) is based on convincing and acceptable elements. The concept of absolute space (and that of absolute time) belongs to such non-acceptable elements.

In fact, Newton's idea of absolute space has led to enormous intellectual problems which — as we will recognize below — could not be eliminated by the Theory of Relativity. What causes the conception of absolute space serious problems? In connection with the term "absolute" two points are relevant:

— Absolute space was invented by Newton for the explanation of *inertia*. However, we do not know any other phenomenon for which absolute space would be responsible. So, the hypothesis of absolute space can only be proved by the phenomenon (inertia) for which it has been introduced. This convenient construction is unsatisfactory and artificial. Science has to eliminate such elements.

— The term "absolute" not only means that space is physically real but also *independent in its physical properties, having a physical effect, but not itself influenced by physical conditions* [5]. Also this feature must be considered as unsatisfactory.

Both points strongly indicate that the concept of absolute space is actually an unphysical quantity. Although Newton's mechanics was very successful — and it is still used in many calculations — a lot of scientists could not accept the concept of an absolute space. This is demonstrated by the fact that physicists tried to solve this serious problem again and again up to the present day. More details are given in Appendix B. (In Appendix B the situation in connection with the General Theory of Relativity is discussed: Within Einstein's theory the absolute character of space-time could not be eliminated; scientists tried it but without success.)

1.3.4 The Situation in General Theory of Relativity

We pointed out above that the effect of "inertia" can only be explained within Newton's mechanics by the "space" in which the body of mass m is embedded whereby space itself has to be considered as an absolute quantity in the sense defined above. The effect of inertia demands that there is a certain relationship between the space and the mass m, i.e., there is an interaction between both elements (space, mass) producing, within the frame of Newton's theory, the effect of inertia. This is however a problematic and ill theoretical construction; such kind of ad hoc conceptions have to be declined.

However, it is interesting to note that the concept of absolute space could even not be avoided within the General Theory of Relativity. The facts are listed in Appendix B. Nevertheless, let us emphasize here one specific fact.

Within Newton's theory a lone body may move through space with constant velocity, indicating the effect of inertia (see Fig. 5b). Exactly the same effect is possible within the General Theory of Relativity. Willem de Sitter demonstrated in the year 1917 that Einstein's field equations lead to the effect of inertia in the case of a lone body moving through space-time, that is, there is exactly that type of inertial motion which is defined within Newton's mechanics.

In a nutshell, *the absoluteness of space, which Newton has claimed, and which Einstein may have attempted to eliminate, is still contained in Einstein's theory* [6].

1.3.5 General Theory of Relativity has a Space-Time Problem

The nature of space and that of time determine the basic "world view". In other words, a wrong idea about space and also about time would influence our basic views about the world seriously. Thus, Newton's theory has to be considered with care because space and time play here the role of "absolute" entities.

However, we have just recognized (Sec. 1.3.4) that also within the General Theory of Relativity there is a problem with space and time, i.e., with the space-time block. This space-time can obviously be the source of inertia, in analogy to that what we have discussed in connection within Newton's mechanics. Further negative details about the strange nature of space-time within the General Theory of Relativity are pointed out in Appendix B.

How could all these unacceptable peculiarities influence the theory itself? How could these strange features be reflected in the concrete results? No doubt, this is difficult to estimate. However, the value for the cosmological constant Λ could be an indication. Before we go in detail, let us first discuss the relationship between the General Theory of Relativity and the conventional quantum theory.

On the relationship between the general theory of relativity and the conventional quantum theory

Physical statements about the cosmos are formulated on the basis of the General Theory of Relativity and also on the conventional quantum theory. Both, the General Theory of Relativity and the usual quantum theory lead to sets of laws that work fantastically. However, if we put them together we inevitably obtain irreconcilable differences. Both theories, also basis for cosmological descriptions, are mutually incompatible.

In Ref. [7] we find: *It has been said that quantum field theory is the most accurate physical theory ever, being accurate to about one part in 10^{11}*, writes Roger Penrose in "The Nature of Space and Time". *However, I would like to point out that general relativity has, in a clear sense, now been tested to be correct to one part in 10^{14} (and this accuracy has apparently been limited merely by the accuracy of clocks on earth).*

In a nutshell, both theories seem to work perfectly but, on the other hand, both theories are mutually exclusive. The estimation of the cosmological constant Λ seems to confirm that.

Estimation of the cosmological constant and cosmic features

The cosmological constant Λ has been introduced by Einstein in a phenomenological way in order to achieve a static universe. However, after Hubble's discovery of an expanding universe, Einstein abandoned and dismissed the concept of a cosmological constant as his "greatest blunder".

Since 1998 we know that the expansion of the universe is accelerated, i.e., the universe expands faster and faster, rather than slowing down as normally expected when we assume that the cosmos is governed by usual matter and gravitation. However, an accelerated expansion needs some sort of antigravitational force, first introduced by Einstein by means of the cosmological constant Λ. In other words, due to the cosmic acceleration there is a need of a nonzero cosmological constant yet.

The cosmological constant can be interpreted as constant pressure which is repulsive for $\Lambda > 0$ and attractive for $\Lambda < 0$. Λ is connected to a certain vacuum energy density, and within the framework of quantum field theory we obtain in fact a non-vanishing energy density. However, such a quantum field theoretical estimation deviates from the observed value by a factor of 10^{122}. This discrepancy in connection with the cosmological constant is often considered as the deepest mystery in physics.

When we connect the quantum field theoretical value for Λ with the theoretical structures of the General Theory of Relativity, we obtain a result which is a catastrophe and we get almost ridiculous cosmic features. In Ref. [8] we find the following instructive example: "Indeed, if the vacuum contained all the energy physicists expect it to, it would be so repulsive that you wouldn't be able to see your hand in front of your nose. Even at the speed of light, the light from your hand wouldn't have time to reach your eyes before the expanding universe pulled it away. *The fact that you can see anything at all, says Krauss, means that the energy of space cannot be so large.*"[8]

However, instead of the conclusion by Krauss we may assume that the quantum field theoretical energy density of space is correct, but not the tenets of the General Theory of Relativity, in particular its space-time conception. In fact, the character of space-time can be absolute in the General Theory of Relativity (here the container principle is valid), and this has to be considered as a serious deficiency of the theory. How does this unacceptable peculiarity influence the theory itself? This is difficult to estimate. (More details in connection with the cosmological constant are pointed out in Appendix C.)

The absolute space (space-time) is the source of inertia, that is, it is able to create physically real effects. Thus, the following question arises: What is the space (space-time) made of? If an entity is able to create physically real effects it should consists of a real something different from matter. In the next sections we will analyse this question. In particular, we will go from the "container principle" to the "projection principle".

1.4 NATURE OF SPACE AND TIME

As we have just found out, the term "absolute" creates essential problems and reflects a non-real situation. A larger problem comes into play when we ask the following question: What is the space (time) made of that is considered as a physically real quantity for the production of inertia? It should be a physically "real something" different from matter. What can we say about this situation?

1.4.1 Are Space and Time Accessible to Empirical Tests?

The basic characterization of space as, for example, our three-dimensional space, is given by points. Each point consists of three real numbers x, y and z that we call coordinates. On the other hand, time is characterized as well by points, and each time point is given by one number, which we have marked above by the Greek letter τ.

If space and time would be physically real quantities, we come to an essential question: Are these basic quantities, i.e., x, y, z and τ, accessible to empirical tests? This is definitely not possible. The following facts show and demonstrate, respectively, that there is no possibility for that.

We definitely cannot see, hear, smell, or taste single elements x, y, z and τ of space and time, that is, the basic elements of space and time, characterized by x, y, z and τ, are not accessible to our senses. This is independent of the character of space (space-time): Whether it is absolute or non-absolute. Also measuring instruments for the experimental determination of the space-time points x, y, z, and τ are not known and even unthinkable.

One might object that we experience space in everyday life (see, for example, Fig. 1). However, this kind of experience has nothing to do with the observation of the basic elements x, y, z and τ. What do we observe in connection with the space-time elements x, y, z and τ? Here the specific facts are relevant: We never observe single elements x, y, z and τ, but we are only able to observe

distances in connection with material bodies (masses),

and

time intervals in connection with physically real processes.

Then, we come to the following important conclusion: Since we are principally not able to "observe" the basic elements of space and time (i.e., x, y, z and τ), space and time should never be the source for physically real effects as, for example, inertia. "Non-observable" here means "non-existent" as a physical and real entity. The scientific standpoint requires such an allocation. This is not fulfilled at all within Newton's theory, but also not within Einstein's theory (Appendix B). The velocity of a lone mass embedded in space (represented in Fig. 5b) remains constant ($\mathbf{v}_1 = const$) and this property is due to a physically real effect: There must be a relationship between Newton's absolute space and the mass m_1, i.e., a certain interaction between both entities (space, mass) has to be postulated, producing, in the opinion of Newton, the effect of inertia. We find similar situations in the Theory of Relativity: Also here a lone body may move through space with constant velocity indicating the effect of inertia (see Fig. 5b).

This is however problematic and is an ill construction because the elements of space-time are not observable with one body alone; as we have already pointed out above, we can only observe "distances in connection with material bodies (masses)", i.e., we need at least two bodies when we would like to make statements about space. Absolute space can therefore not be considered as a physically real something. Such a construction has to be declined.

As we have already remarked, similar ill situations come into play within the General Theory of Relativity; this is quoted and discussed above and in particular in Appendix B. Within the Special Theory of Relativity we have a situation similar to that in Newton's theory. Here the absolute space of Newton is merely replaced by an absolute space-time.

In summary, the basic elements of space and time x, y, z and τ do not reflect physically real quantities and cannot be the source of physically real effects. From this point of view, an empty space-time should not exist because it is not observable and, from the point of view of science, only those entities which can lead to physically real effects are observable. Nobody can cut out a piece of space or a piece of time and is able to put them on the table. Nobody can invent an experimental procedure for that. In fact, such a possibility reflects a

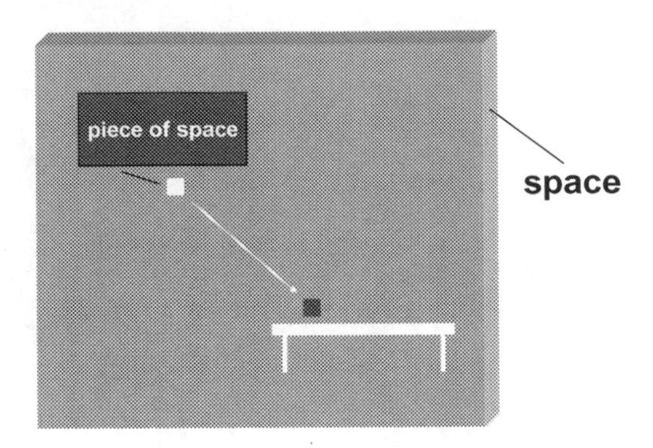

Fig. 6 What is the space made of? Can we cut out a piece of space and can we put this piece on a table? Nobody is able to do that, and nobody can invent a procedure for that. The question is simply ridiculous.

ridiculous situation, and this is because no-one can imagine that (see also the illustration in Fig. 6).

Clearly, the basic elements of space and time, i.e., x, y, z and τ, cannot be identified with a "real something" in analogy to matter.

1.4.2 Ernst Mach

Ernst Mach (1838–1916) rejected Newton's absolute space radically. Also Mach based this rejection on the fact that the basic elements of space and time, characterized by x, y, z and τ, are not accessible to empirical tests. Therefore, Mach argued that these quantities cannot form a "physically real something" to be capable of physically real effects such as, for example, the effect of inertia.

In other words, according to Mach, the space (space-time) can never be the source for physically real effects, that is, the space (space-time) can never act on material objects giving them certain properties (inertia). According to Mach, a particle does not move in un-accelerated motion relative to space, but relative to the center of all the other masses in the universe. This statement reflects Mach's principle often discussed in literature. Mach's principle is entirely based on the above discussed fact that we can never observe that what we call space (space-time) because its elements (coordinates x, y, z and time τ) are in principle not observable. We can only say something about distances in connection with masses, and time intervals in connection with physical processes.

Again, space and time can never be the source for physically real effects, i.e., the space-time block is not a physically real entity like matter. However, Mach's principle goes not far enough because this principle does not explicitly forbid matter to be embedded within the space-time block. In this monograph we will use the term "Mach's principle" in connection with the fact that space and time can never be the source of physically real processes. As far as we know, Mach never forbade the possibility for matter to be embedded in space although this peculiarity should follow from Mach's principle. But how can a physically real something (matter) be embedded within a system that is not a physically real entity (space-time)? This is hardly possible. Such a conception may be implied when we base our direct knowledge on that what we have directly in front of us (Fig. 1) but, as we have recognized, these everyday life impressions do not reflect reality itself but "only" an image of it. Because a physically real something (matter) can hardly be embedded within a system that is not a physically real entity (space-time), we propose in this monograph to replace the "container principle" by another, more realistic physical principle, namely the "projection principle".

Remark

Mach's statements are based on Newton's mechanics, in particular on the nature of space used within the framework of Newton's theory. We have to answer the following question: How can we understand Newton's absolute space (time) which can even exist without the existence of material bodies? Sure, it is by definition the seat of inertia. But what is it? What is it made of?

As we found out, it is made of nothing because it cannot be the source of physically real effects. This in particular means that the elements of space and time, characterized by x, y, z and τ, are not observable. Such a space-time block cannot have any physically real existence. It is nothing! This kind of construct cannot be taken as the foundation of a scientific theory, in particular in connection with the theory of motion where the masses are embedded in space and time. Newton closed this logical gap by associating the empty space (time) with divine attributes. So Newton combined — against his own principles — physical with metaphysical arguments.

Since Galileo Galilei the tendency is predominantly not to use dogma (that is, a system of beliefs, put forward by some authority to be accepted without scrutiny) in the scientific treatment of real phenomena in nature. One reason why Newton's absolute space and absolute time are dogmatic in character is that they may possibly reflect the divine omnipresence.

Therefore, Newton's conception of space and time has been criticized immediately by influential philosophers Berkeley and Leibnitz. But the big success of Newton's mechanics overshadowed these serious doubts.

1.4.3 Einstein and Mach's Principle

Albert Einstein was strongly influenced by Mach's ideas and created, as we know, new space-time theories. Nevertheless, he could not completely eliminate the absoluteness of space (space-time). As we have pointed out above and in Appendix B, within the General Theory of Relativity as well as within the Special Theory of Relativity, the absolute character of space-time could not be completely eliminated. This is particularly reflected within the Special Theory of Relativity where we have an absolute four-dimensional space-time which has exactly the same character as Newton's absolute three-dimensional space.

Mach strongly needed to eliminate space (space-time) as an active cause. According to him there should be no physically real effects due to space (space-time) as, for example, the effect of inertia. According to Mach, a material body does not move in un-accelerated motion relative to space, but relative to the center of all the other masses in the universe. Note that Mach argued explicitly on the basis of the container principle.

1.4.4 Some Basic Statements

The basic elements of space and time, characterized by x, y, z and τ, are not accessible to our senses. We definitely cannot see, hear, smell, or taste them. Also measuring instruments for the experimental determination of the space-time points x, y, z, and τ are not known and even unthinkable. As we have already outlined above, we can only say something about "distances in connection with masses", and "time intervals in connection with physical processes". What are the consequences of these facts?

Formal analysis

What does the statement "we can only say something about distances in connection with masses" mean? We want to investigate this point in more detail.

Let us consider two bodies 1 and 2, having the masses m_1 and m_2 and the coordinates x_1, y_1, z_1 and x_2, y_2, z_2 at time τ. "Matter" and "space" are closely linked; neither should be able to exist without the other. This point is important and needs further explanations.

The space-points x_1, y_1, z_1 and x_2, y_2, z_2 are connected to the masses m_1 and m_2. That is, we always have the relations

$$\begin{aligned} &x_1(m_1),\ y_1(m_1),\ z_1(m_1) \\ &x_2(m_2),\ y_2(m_2),\ z_2(m_2) \end{aligned} \qquad (1)$$

The point is however that we never register an isolated point, i.e.,

$$x_1(m_1),\ y_1(m_1),\ z_1(m_1)$$

or

$$x_2(m_2),\ y_2(m_2),\ z_2(m_2),$$

but in any case only the distances:

$$\begin{aligned} &x_1(m_1) - x_2(m_2) \\ &y_1(m_1) - y_2(m_2) \\ &z_1(m_1) - z_2(m_2). \end{aligned} \qquad (2)$$

Our basic statement that we can only observe "distances in connection with masses" is expressed mathematically by Eq. (2).

The elements x_1, y_1, z_1 in isolated form are not accessible to empirical tests, but only in connection with the elements x_2, y_2, z_2. Clearly, the elements x_2, y_2, z_2 in isolated form are also not accessible to empirical tests, but only in connection with x_1, y_1, z_1. At first, we have to accept this feature as a matter of fact; it reflects a specific peculiarity of space and is connected to the question whether the space is a physically real entity or not. We will analyze this point further; in particular, we will find out below that the elements x_1, y_1, z_1 and x_2, y_2, z_2 somehow cannot be considered hidden quantities, but they are actually not definable. In other words, the elements x_1, y_1, z_1 and x_2, y_2, z_2 in isolated form are not only not observable but they are simple not defined, and the "definition" of a feature is the presupposition for its observation. In Sec. 1.4.7 we will analyze the physical background of these facts in more detail.

Matter and space

Matter and space are closely linked; neither should be able to exist without the other. The reason is simple: Our observations are always done within space. Matter is assumed to be embedded in space, and we never observe that matter leaves space or enter it. What is a body of a certain mass without space? This is not imaginable and it makes obviously no sense. For example, Newton's equations of motion would not be defined. That is, a mass without space would make not much sense, when we base our knowledge on the concepts of conventional physics.

This in particular means the following: The space coordinates x_1, y_1, z_1 and x_2, y_2, z_2 can only exist in connection with the masses m_1 and m_2 (Eq. (1)); the masses m_1 and m_2 can only exist in connection with space coordinates, i.e., in our example we need x_1, y_1, z_1 and x_2, y_2, z_2. In other words, besides Eq. (1) we must have the following relations:

$$\begin{aligned} &m_1(x_1),\ m_1(y_1),\ m_1(z_1) \\ &m_2(x_2),\ m_2(y_2),\ m_2(z_2). \end{aligned} \tag{3}$$

Since the elements x_1, y_1, z_1 cannot exist in isolated form, it follows also that the mass m_1 cannot exist as a lone entity. This relationship holds for x_2, y_2, z_2 and m_2 as well.

Thus, we may give the following general statement: Because isolated space-positions are not existent, a body cannot be defined relative to space, but only relative to another body.

1.4.5 Consequences and Illustrations

Equation (3) (and also Eq. (1)) has to be considered as the "minimum information" for the observation of material bodies in the world. This in particular means that the mass m_1 or m_2 cannot exist in isolated form, that is, an isolated single mass as for example m_1 cannot exist in space, but at least two masses are needed in nature, i.e., m_1 and m_2. We always observe or measure distances (Eq. (2)) and for this purpose we need two bodies.

This situation is illustrated in Fig.7. In this figure a single, isolated elementary body (body A), which is characterized by a mass m_A and a geometrical point with the coordinates x_A, y_A, z_A, cannot be observed and is therefore not a physically real entity. Here "elementary body" means a body without inner structure (without subsystems).

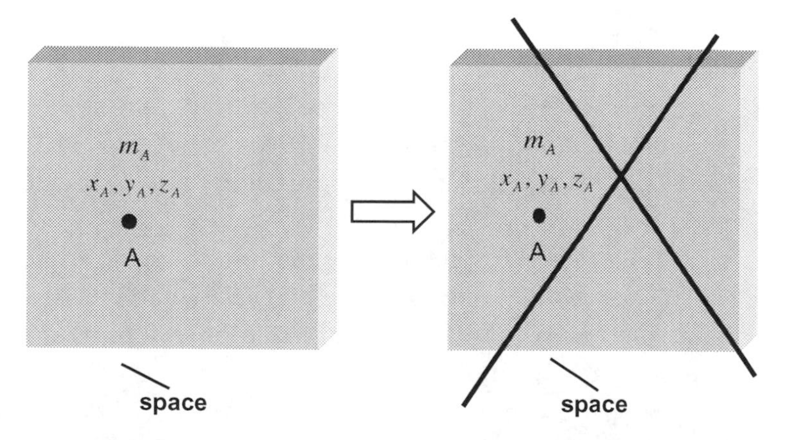

Fig. 7 Only one elementary body (body A), having the mass m_A and the space position x_A, y_A, z_A, is embedded in space. Such a situation cannot have any physically real existence. Such a configuration does not fulfil the minimum information (see Eq. (1)), which is necessary for observations. Therefore we have to reject this construction.

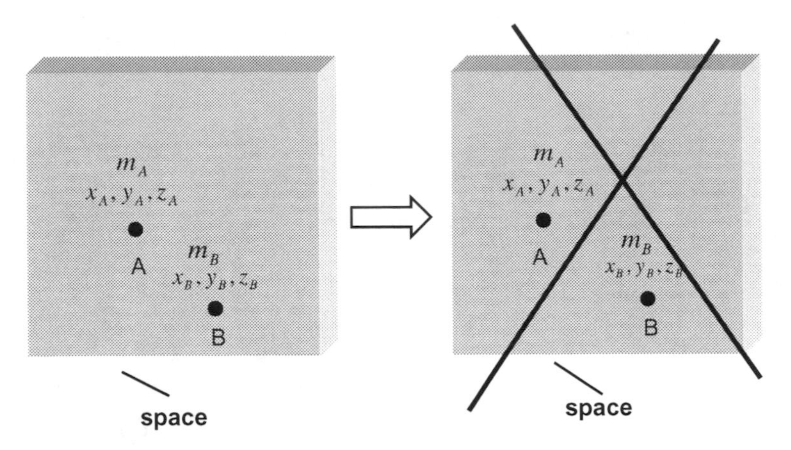

Fig. 8 In this figure we go a step further and put two elementary bodies (body A and body B) into space; body B is characterized by a mass m_B and a geometrical point with the coordinates x_B, y_B, z_B and body A again by the mass m_A and the coordinates x_A, y_A, z_A. We would like to assume that there is no interaction between the two bodies with m_A and m_B. If there is no interaction between the two bodies with m_A and m_B, body A is not existent for body B, and body B is not existent for body A. Then, body A remains an isolated non-existing entity and also body B must be classified as an isolated non-existing entity. In other words, body A and body B are not defined and, therefore, we are not able to define a distance between body A and body B. An empty space remains. As in the case of Fig. 7, we have also here an empty space, characterized by an infinite ensemble of coordinates x_i, y_i, z_i, which cannot exist without any material body since it does not fulfil the minimum information Eq. (1). In conclusion, this configuration also makes no sense and must be excluded.

In other words, the elementary body in Fig. 7 does not exist and an empty space remains. However, an empty space, characterized by an infinite ensemble of coordinates x_i, y_i, z_i, also cannot possibly exist, because it does not fulfil the minimum information Eq. (1). Evidently, Fig. 7 makes no sense.

In Fig. 8 we go a step further and we put into the same space two elementary bodies (body A and body B); body B is characterized by a mass m_B and a geometrical point with the coordinates x_B, y_B, z_B and body A by the mass m_A and the coordinates x_A, y_A, z_A. We would like to assume that there is no interaction between the two bodies with masses m_A and m_B. Then, body A is not existent for body B, and vice versa body B is not existent for body A. In other words, body A behaves like a unit alone in space; the same is true for body B:

Behaving like a unit alone in space. Therefore, both bodies may not exist, as in the case of a lone body in space (Fig. 7).

Clearly, body A and body B are isolated entities, which do not exist. What we have said in connection with a lone particle in space is also valid for the configuration consisting of body A and body B: Both bodies A and B in Fig. 8 are isolated and are therefore non-existing entities. Because body A and body B are not defined, we are not able to define a distance between body A and body B.

As in the case of Fig. 7 an empty space remains. However, here an empty space, characterized by an infinite ensemble of coordinates x_i, y_i, z_i, cannot exist since it does not fulfil the minimum information Eq. (1). Evidently, it follows also that Fig. 8 makes no sense and must be excluded.

In summary, in order to fulfil the minimum information (Eq. (1)), we not only need the space coordinates

$$x_A, y_A, z_A, \quad x_B, y_B, z_B$$

and the masses

$$m_A \quad \text{and} \quad m_B$$

of the two bodies, but there must be in addition a "relation" between them in order to avoid that scenario where A and B are isolated. Such a relation is generally expressed by an interaction between body A and body B, and this situation is represented in Fig. 9; due to this interaction both bodies are able to exist and a distance between them can be defined.

Again, A and B cannot exist as free, non-interacting bodies, but we need an interaction between them which is "existence-inducing". What is the physical background of this interaction type?

1.4.6 Existence-Inducing Interactions

Only such kind of systems, represented in Fig. 9, are able to exist, because only such kind of systems can be defined, and — as we already remarked — the definition of a property is the presupposition for its observation.

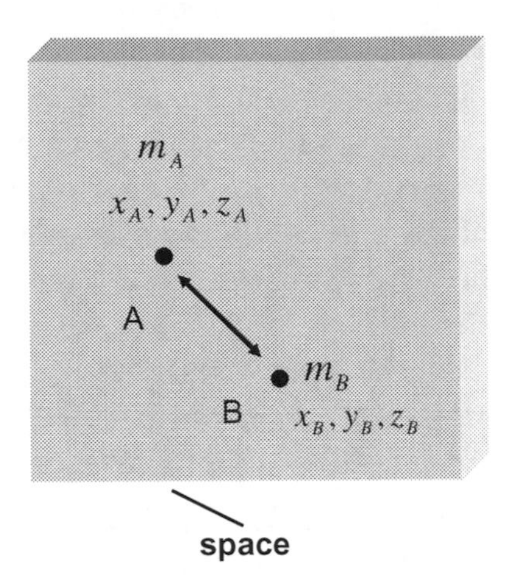

Fig. 9 In this figure we have the same configuration as in Fig. 8. However, we have now an interaction between the two bodies A and B; it is an "existence-inducing interaction". Only in this way the configuration can come into existence. Then, both bodies exist and a distance between them can be defined. Only such a kind of configuration is able to fulfil what is observable in nature and what is scientific.

Only such system-configurations fulfil what is observable in nature and what is scientific: We can only say something about "distances in connection with masses" (and time intervals in connection with physical processes). However, this statement is based on the conception that the material world is embedded in space (space-time) when, in other words, the situation represented in Fig. 3 is applied. We will recognize below that this "container principle" should not reflect a physically real situation. Nevertheless, we will also recognize that the analysis, which we have made here, is still valid when we apply it on the situation given in Fig. 2: there are geometrical positions instead of the real masses.

Let us come back to what we have called above "interaction". It is, as we have pointed out, necessary for the construction of a realistic physical reality. What kind of interaction is required? It must be an interaction that is not in effect between already existing elementary bodies, but it is an interaction which begets and produces the elementary bodies itself. In other words, body *A* produces body *B* and body *B* produces body *A*. Such an "existence-inducing" interaction is necessary because body A and body B cannot exist as free, non-interacting systems (Sec. 1.4.4). As a matter of fact, we need body A and body B for the definition of space-distances (see Eq. (2)).

The existence-inducing interaction should be independent of the distance between the bodies because the existence of a body with definite properties (for example its mass) should be independent of the distance between the bodies. When both bodies are produced by this existence-inducing interaction, they may in addition interact via a distance-dependent pair interaction. However, for fulfilling the minimum information Eq. (1) only distance-independent "existence-inducing" interactions are relevant and necessary.

All distances have to be considered as equivalent. Therefore, the notion "distance" should be independent of the interaction between the masses m_A and m_B of the bodies A and B, and this is given when the existence-inducing interaction is distance-independent. In conventional physics the construction of a certain distance-dependent interaction needs a certain frame, which allows us to define distances, as presupposition for the construction of the interaction law so to say.

1.4.7 Summary and Further Statements

Before we give some further relevant statements, let us summarize the main facts:

1. Isolated space-points are not defined and, therefore, not accessible to empirical tests. The definition of a certain property is the presupposition for observations. The space-elements, say x_A, y_A, z_A, are only definable in connection with a second space-point, say x_B, y_B, z_B. Furthermore, each of these space-positions must be coupled to a body having a certain mass. We have coupled x_A, y_A, z_A by m_A and x_B, y_B, z_B by m_B.
2. The elements x_A, y_A, z_A and x_B, y_B, z_B can be chosen arbitrarily but they must fulfil relation Eq. (2), i.e. only the distances $x_A - x_B, y_A - y_B, z_A - z_B$ are physically relevant.
3. There must be a "relation" between the masses m_A and m_B, i.e., there must be an interaction between them, which are

existence-inducing. These kind of interaction should be distance-independent.

4. We may give the following general statement: Because isolated space-positions are not existent, a body cannot be defined relative to space, but only relative to another body.

Further statements

Distances between bodies can only be defined if there exists an existence-inducing interaction between the masses of the bodies. This interaction, for example between the masses m_A and m_B, produces a *relation* between the space-coordinates x_A, y_A, z_A and x_B, y_B, z_B, and this relation can be expressed mathematically by (see Eq. (2)):

$$x_A(m_A) - x_B(m_B)$$
$$y_A(m_A) - y_B(m_B)$$
$$z_A(m_A) - z_B(m_{B2})$$

This suggests that the "relation" (correlation) between the space-points x_A, y_A, z_A and x_B, y_B, z_B are produced by an interaction between these space-points, i.e., between x_A, y_A, z_A and x_B, y_B, z_B, instead of an interaction between the masses m_A and m_B. However, such kind of space-space interaction is not conceivable. We cannot put a piece of space on the table (see in particular the discussion in connection with Fig. 6). Then, the relation between the space-points x_A, y_A, z_A and x_B, y_B, z_B comes entirely into existence through the interaction between m_A and m_B.

Why can an isolated space-point as, for example, x_A, y_A, z_A not be defined? The answer is simple. There does not exist a space marking, i.e., the space cannot be labelled with definite numbers. In other words, we cannot allocate a certain space-position (for example x_A, y_A, z_A) to an isolated body. A position relative to space is not defined for an isolated body, but only a position relative to another space-point (x_B, y_B, z_B). Thus, a body, say A, cannot be defined relative to space, but only relative to another body B.

This is the reason why the space (space-time) cannot be judged as a physically real entity. In fact, we cannot put a "piece of space" on the table, and also not a "piece of time". The physical space (space-time) cannot be identified with a certain kind of substratum; there is no indication for that. The space-time is not made of a real something (substratum) in analogy to matter.

The elements x_A, y_A, z_A or x_B, y_B, z_B in isolated form are not defined and, therefore, they are not accessible to empirical tests, but only in connection with the elements x_B, y_B, z_B or x_A, y_A, z_A. Only the distances between x_A, y_A, z_A and x_B, y_B, z_B are relevant. Then, the elements x_A, y_A, z_A and x_B, y_B, z_B can be chosen *arbitrarily* but they are not independent from each other because they must fulfil relation Eq. (2), i.e., we have $x_A - x_B, y_A - y_B, z_A - z_B$. This reflects the peculiarity that there should be no interaction between space-points. If the space would reflect a physically real entity, the world around us could not have the properties we experience.

1.4.8 Conventional Physics versus Projection Theory

Within conventional physics (Newton's mechanics, quantum theory, the Theory of Relativity, etc.), where we work within the "container principle", distance-independent "existence-inducing interactions" are not known and would be difficult to introduce. The world is however not embedded in space and time within the projection theory ([1, 2], Appendix F). In this case, all the points in connection with inertia, system-producing properties etc. can be treated without any problem. For example, isolated, non-interacting elementary systems are forbidden to exist in the projection theory, and this is the outcome of the theory. Just our basic statement about that what is observable (we can only observe "distances in connection with masses") requires that isolated elementary bodies may not exist. This is definitely not fulfilled within the frame of Newton's mechanics and also not in connection with the other theories of conventional physics, but within "projection theory". More details will be given below.

Within Newton's mechanics non-interacting bodies may exist in space; this situation is represented in Fig. 5a. Here the bodies move through space with constant velocity. In this case the minimum information expressed by Eq. (1) is definable, and we can formulate distances between the non-interacting bodies.

In Newton's mechanics there is a correlation between the bodies, and this correlation is given by the postulated inertia produced by space. Thus, we may speak of an indirect interaction between the bodies which move through space with constant velocity. In Newton's mechanics we may define isolated elementary bodies and we may put them into space, i.e., the world is embedded in space. However, as we have remarked several times, the postulated concept of inertia is an unphysical concept. Similar ill constructions can be found within the two forms of the Theory of Relativity.

1.5 INSIDE WORLD AND OUTSIDE WORLD

In the preceding sections we have discussed the question of how traditional physics treats space and time. The nature of inertia has been analyzed within the frame of Newton's mechanics. Here space is the source of inertia, i.e., space produces inertia. In other words, space is able to produce physically real effects and, therefore, within Newton's mechanics space and time have to be considered as physically real entities. This is however an ill construction because the basic elements of space and time, i.e., x, y, z and τ, are not observable. Then, from the point of view of science, space and time cannot represent physically real quantities.

We definitely cannot see, hear, smell, or taste single elements x, y, z and τ of space and time; that is, the basic elements of space and time, characterized by x, y, z and τ, are not accessible to our senses. This is a general statement and is independent of whether space and time are absolute or non-absolute. Also specific measuring instruments for the experimental determination of the space-time points x, y, z, and τ could not be invented and is even unthinkable. The only physical property on which we can base our knowledge about space and time is given by the following fact: We can only say something about "distances in connection with masses", and "time intervals in connection with physical processes".

To sum up, we are principally not able to "observe" the basic elements of space and time (i.e., x, y, z and τ); therefore, from the point of view of science, space and time should never be the source for physically real effects as, for example, inertia.

Nevertheless, the phenomenon of space and that of time is existent. Everybody knows intuitively what space (time) is. Our world in front of us appears as a space-container filled with real objects that we feel and observe. The same is true for time: We all believe to know what time is but, however, when we try to understand the "nature of time" we normally evade this question.

In other words, we feel and observe real objects, but the basic elements of the container, the space-time points x, y, z, and τ, in which these objects are embedded, may not to be considered as physically real. No doubt, this is somehow a contradiction. In this section we would like to propose a solution for this problem.

1.5.1 Are Space and Time Elements of the Outside World?

In Fig. 1 a typical configuration is given, which we have, in everyday life, spontaneously in front of us. When we touch with our fingers an object as, for example, one of the trees in Fig. 1, we definitely feel it, i.e., the tree and the observer's body interact with each other. Both, the tree as well as the observer, are considered as physically real objects. This is in a certain sense correct, but we have to be careful. Do we have really the material reality directly in front of us? No, we have not. Already a lot of scientists, in particular the philosopher Immanuel Kant pointed out this fact (see also Sec. 1.2).

Let us repeat the statement by C.G. Jung (see Sec. 1.1.1): *When one thinks about what consciousness really is, one is deeply impressed of the wonderful fact that an event that takes place in the cosmos outside, produces an inner picture, that the event also takes place inside ….* [3]

In other words, in the opinion of C.G. Jung we have an "outside world" and we have simultaneously an "inside world". What does it mean in detail? When we touch with our fingers certain objects (tree, car, etc.) we definitely feel them, i.e., the objects and the observer's body interact with each other. Both, the objects as well as the observer, are considered as physically real objects. We make this statement on the basis of the facts of the inside world, but there are no material objects within this inside world. This is however no problem because it is normally assumed that there is an exact "one-to-one-correspondence" between the reality outside and the inner picture of it. This concept is also used in traditional physics, in particular within Newton's mechanics. We will recognize below that this assumption is more than doubtful; a lot of facts speak against this concept, and some of them will be quoted below.

Nevertheless, C.G. Jung's statement with respect to "outside world" and "inside world" is correct, but the assumption that an event takes place twice is obviously not realistic. As we have already remarked in connection with Fig. 2 and Fig. 3, we may not assume that there is a "one-to-one correspondence" between the reality outside and the inner picture of it. This is far from being realistic. The reason is obvious and will be analyzed below.

First, let us answer the following question: Are space and time with their basic elements x, y, z, and τ really entities of the outside world? No doubt, space and time are elements of the inner world, and this fact cannot be denied because we experience them within the frame of assumption-less observations in everyday life (Fig. 1 is a typical example of such a situation).

Assumption-less observations are the most direct experiences we have from the world outside. Therefore, if a theory violates a result of an assumption-less observation, this theory has to be rejected. Clearly, this is also the case when we study the facts of the world by means of certain measurements; a measuring instrument is always constructed on the basis of a certain theoretical world view.

In Fig. 1 we do not have objects that are embedded in space and time having the elements x, y, z, and τ. We merely observe "objects" and "extensions". For example, two objects have a certain "extension". Here "extension" has to be considered as a basic notion and we should not try to analyze it any further, i.e., we should not try to explain the notion "extension" by more basic notions, more basic than the notion "extension" itself. "Extension" reflects a qualitative effect, and we actually meet it first in its basic form just in connection with our assumption-less observations in everyday life. The effect of "extension" appears spontaneously without thinking and, therefore, it has to be classified as a qualitative element. This assumption-less space-time-impression always reflects an inner image and is positioned within the brain of the observer (an example is represented in Fig. 2). Clearly, also Fig. 1 has to be classified as an inner image in analogy to Fig. 2.

We have a lot of bodies in front of us and, therefore, we have a lot of extensions. The brain organizes this "ensemble of extensions" as *one* phenomenon which we call "space". That is all what we can say about space within our assumption-less everyday impressions. Insofar space is a qualitative impression of human observers. But we need more than qualitative impressions when we want to analyze the physical processes in the world theoretically.

Within the frame of theoretical considerations we can hardly work on the basis of this qualitative space-impression, but we need a more sophisticated conception. For quantitative descriptions the observer normally works on the basis of a three-dimensional space, and this is in accord with our impressions in everyday life. In other words, for the quantitative description of certain three-dimensional space configurations at time τ — as for example those that are given in Fig. 2 and Fig. 1 — we use three coordinates x, y, z that allow us to position the geometrical points relative to each other on the basis of three numbers. Then, for the quantitative description of an "extension" of two geometrical points (two bodies A and B in the world outside having the masses m_A and m_B) we need the information $x_A, y_A, z_A, x_B, y_B, z_B$.

In essence, we have always "extensions" which are characterized within a quantitative description by distances. We need at least two bodies for the direct observation of extensions. In the case of two bodies we use six numbers for the determination of the distance between them. This is the minimum information (see Sec. 1.4.4); note that there cannot be real masses m_A and m_B in the image but merely geometrical positions and, furthermore, there cannot be interactions between geometrical positions.

We have noted in Sec. 1.4.4 that the masses m_A and m_B belong to the minimum information and a certain kind of interaction between the bodies. However, this refers to the world outside (Fig. 9). This minimum information can however be transferred to the situation in the image: Here the material bodies, having the masses m_A and m_B, are replaced in the image by the geometrical positions A and B, but this situation does not affect the statement with respect to the minimum information.

Due to the exchange of m_A and m_B by A and B, the scenario takes a somewhat other form: The observer is not only positioned as a material object in the world outside, but he simultaneously appears as a geometrical object in the image; he feels the effects due to the masses m_A and m_B (their real interaction with the observer) when he touches the geometrical positions A and B in the image. There is a definite "correlation" between the observer's touch in the image and the real effect in the world outside. Essentially, the observer feels the real bodies with the masses m_A and m_B at the geometrical positions A and B in the image that we have in front of us. The "minimum information" is fulfilled when we simply exchange "interaction effects" in the world outside by "correlations" in the inside world; that is, when we work with correlations within the image which is directly in front of us in everyday life observations.

In the formulation of the minimum information (Sec. 1.4.4) a real interaction (existence-inducing interaction) between the masses m_A and m_B in the "world outside" is necessary. This real existence-inducing interaction between these physically real masses becomes a strict "correlation" between the geometrical points A and B in the image. The interaction itself takes place in the world outside where the real masses are positioned.

To summarize, the whole scenario in connection with that what we have called "minimum information" can be transferred from the world outside to the inside world (identical with the image in front of us) where only geometrical configurations are positioned. More details will be discussed below.

The coordinates x, y, z at time τ are here the elements of a "fictitious net" which the observer intellectually puts over the image in front of him. This process is indicated in Fig. 10. That is all what we can say about the space-time elements x, y, z and τ because more far-reaching statements about the basic elements of space and time cannot be proved experimentally, and such statements would not be scientific, as we outlined above several times. This in particular means that the basic elements x, y, z and τ are entities of the brain because they come into existence by thinking: The observer intellectually puts x, y, z and τ over the image in front of him.

In a nutshell, the entities x, y, z and τ have nothing to do with a "real something". Let us deepen this point.

1.5.2 World Outside within the Container Principle

What about the existence of the basic space-time elements x, y, z and τ in the world outside? Does Fig. 3 really reflect a realistic configuration? May we really assume that the container principle can be applied in connection with the outside world? It is valid within Newton's mechanics and all the other approaches in traditional physics. However, as we have already outlined above, we have to be careful since we are not able to observe the coordinates x, y, z and the time τ. This means, from the scientific point of view, that the quantities x, y, z and τ cannot be considered as a "real something" analogous to the material bodies. It is unthinkable that the material world outside is embedded within such a kind of container, which consists of nothing. In other words, the container principle is a doubtful concept which has to be eliminated. This statement is relevant when we discuss the structure of the world outside, and we will justify this position by further arguments.

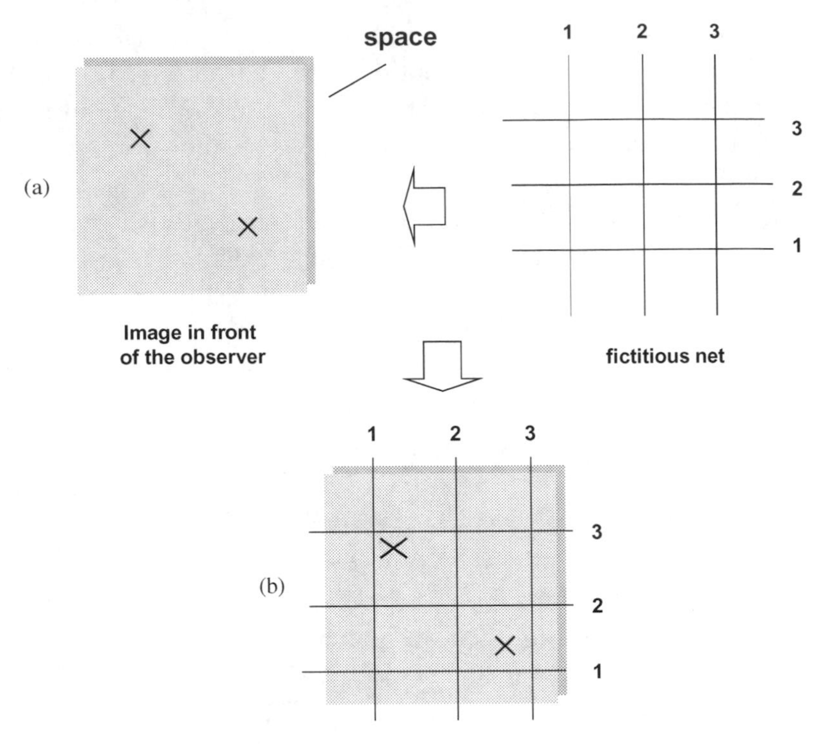

Fig. 10 (a) This figure is identical with Fig. 2; it reflects an inner image and is positioned within the brain of the observer. We normally have such images in front of us within assumption-less observations in everyday life. The two crosses in the figure are the geometrical positions of two material bodies which belong to the outside world (Fig. 3). Also the more complex space-configuration in Fig. 1 reflects such a geometrical situation at a certain time τ; it is an inner image which is positioned within the brain of the observer. Such figures (Fig. 2, Fig. 1) are qualitative in character and consist of "geometrical positions" and "extensions". Within the frame of theoretical considerations we can hardly work on the basis of such a qualitative space-impression, but we need a more sophisticated conception. For quantitative descriptions the observer normally works on the basis of a three-dimensional space, and this is in accord with our impressions in everyday life. For a certain three-dimensional space-configuration at time τ we use three coordinates x, y, z that allow us to position the geometrical points relative to each other on the basis of three numbers. The coordinates x, y, z at time τ are here the elements of a "fictitious net" which the observer intellectually puts over the image in front of him and we come to Fig. 10b, this figure represents a two-dimensional situation.

We stated above that the basic elements x, y, z and τ are entities of the brain because they come exclusively into existence by thinking. We need the quantities x, y, z and τ when we want to "change" reality in front of us (examples are given in Fig. 1 and Fig. 2). Here certain distances between bodies are often relevant, and such distances are only defined and determined by means of the positions of the bodies; each of these positions are expressed at time τ by three numbers, that is, by specific values of x, y, z.

Clearly, when we merely want to "observe" the world in front of us we do not need a quantitative description on the basis of x, y, z and τ, but only a qualitative perception which is relevant, which is given by the geometrical image of the bodies and their "extension". Here the notion "extension" reflects, in accordance with our definition above, a qualitative impression (Sec. 1.5.1).

In the case of the container principle (realized, for example, in Fig. 3) the real material bodies are embedded in space and time, and there are of course no geometrical positions. This situation reflects the world outside within the container principle. The world outside is not static, but there are dynamic and evolutionary changes, and these processes take place without specific actions of the observers. The execution of such changes, done by nature itself, provides the minimum information about space and time. To fulfill this minimum information it is not sufficient that nature has only a qualitative impression of space as is expressed by the notion "extension", but a more sophisticated space-time knowledge in the form of the basic elements x, y, z and τ would be necessary. Then, nature could work with relative distances etc. In essence, within the container principle (Fig. 3) the entities x, y, z should be existent at each time τ intuitively without thinking. This is trivial because the world outside is assumed to be existent without an observer. Since the container, made of space and time, is filled with matter, the container itself must be a real something within the frame of this conception. But we do not observe the elements x, y, z and τ! Thus, also from this point of view the container principle reflects an ill, non-scientific conception.

We noted above, that in the case of the inside world the basic elements x, y, z and τ come exclusively into existence by thinking. These inner worlds are represented for example by Fig. 1 and Fig. 2. No doubts come up when we analyze the configurations of the inner world (the image of the world outside) with the quantities x, y, z

and τ. In other words, no problems appear when the coordinates x, y, z at time τ are the elements of a "fictitious net" which the observer intellectually puts over the image in front of him, as indicated by the construction in Fig. 10.

1.5.3 Isolated Coordinates

However, serious problems come up when we assume that the basic elements x, y, z and τ are also existent in the world outside as in the case of the container principle. Here space and time have to be considered as physically real quantities and could be, for example, the source of inertia. This is however a more than doubtful conception as we have recognized above in connection with Newton's mechanics. There is another serious problem if we assume that the container principle is correct, i.e., when Fig. 2 is the image of Fig. 3. The inner world (Fig. 2) reflects the facts of the outside world (Fig. 3). How is the information, which belongs to the world outside, transferred to the inner world? The entire complex, consisting of two material bodies and the distance between them, must appear in the inner world as an image. This question must be answered when we work within the framework of the container principle.

Clearly, the two material bodies with m_A and m_B appear in the image as geometrical positions A and B, but the two geometrical positions, expressed at time τ by the coordinates $x_A, y_A, z_A, x_B, y_B, z_B$, must also be transferred from the world outside to the inner world. Otherwise we have no information in the image about the distance between the two geometrical positions A and B.

This is perhaps no problem because the coordinates x_A, y_A, z_A and x_B, y_B, z_B appear together with geometrical positions A and B, i.e., our basic statement formulated in Sec. 1.4.1 is fulfilled when we replace the statement "*at time τ we can only say something about distances in connection with masses*" by the equivalent formulation "*at time τ we can only say something about distances in connection with*

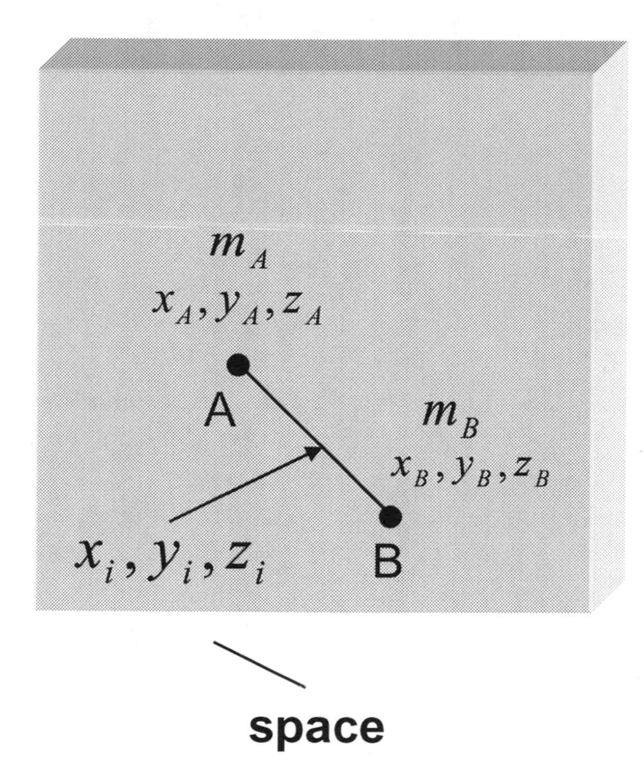

Fig. 11 It is assumed that the container principle is valid. The world outside consists of a container which is filled with two material bodies having the masses m_A and m_B and the coordinates x_A, y_A, z_A and x_B, y_B, z_B at time τ. In other words, for the positions between the coordinates x_A, y_A, z_A and x_B, y_B, z_B no further material objects are defined, but there are isolated coordinates x_i, y_i, z_i. However, isolated coordinates like x_i, y_i, z_i are principally not observable and cannot have any real existence within the world outside.

geometrical positions". That is, the minimum information Eq. (1) is fulfilled for geometrical positions.

This is however not fulfilled for the coordinates x_i, y_i, z_i between the two material bodies with m_A and m_B (see Fig. 11), and this is because there are no other material bodies between the two material bodies A and B. As we already know from our discussion above, isolated coordinates like

$$x_i, y_i, z_i$$

with

$$x_A > x_i > x_B,$$
$$y_A > y_i > y_B, \qquad (4)$$
$$z_A > z_i > z_B$$

are principally not observable at time τ; they do not fulfill the minimum information Eq. (1) and, therefore, they cannot have any real existence within the world outside because only the properties of a physically real something can be transferred from the world outside to the inner world. On the other hand, the existence of x_A, y_A, z_A and x_B, y_B, z_B requires that the isolated coordinates x_i, y_i, z_i between x_A, y_A, z_A and x_B, y_B, z_B are existent.

If space and time would actually exist outside we could assume that the coordinates x_A, y_A, z_A and x_B, y_B, z_B are existing quantities because they fulfil the minimum information Eq. (1). This would however not be the case for the isolated coordinates x_i, y_i, z_i between x_A, y_A, z_A and x_B, y_B, z_B (defined by Eq. (4)). However, as we know, the coordinates x_A, y_A, z_A and x_B, y_B, z_B cannot exist outside because there can be no space and time in the world outside.

Clearly, space-time and its basic elements x, y, z and τ can only appear in the inner world and they do not belong to the elements of the world outside. That is, the container principle is obviously an unrealistic conception.

The coordinates x, y, z at time τ are exclusively the elements of a "fictitious net" which the observer intellectually puts over the image in front of him. Thus, space and time, constructed in this way, can never be the source of physically real effects. That is all what we can say about the space-time elements x, y, z and τ. This in particular means that the basic elements x, y, z and τ are entities of the brain because they come into existence by thought. Then, the coordinates x_i, y_i, z_i at time τ between the positions x_A, y_A, z_A and x_B, y_B, z_B are automatically defined by the "fictitious net" which the observer intellectually puts over the image. The situation is illustrated in Fig. 10.

Remark

We never "observe" isolated space-time positions characterized by x, y, z and τ. At time τ we can only observe "distances" in connection with geometrical positions (Sec. 1.4.1), i.e., we need at least two bodies. We never observe the space-time positions of an empty space or the space-position of only one body; such situations are conceivable but not observable.

There is no physically real process in the world outside that would lead to an empty space or a space with only one geometrical position in the image, corresponding to a non-interacting elementary body in the world outside (see in particular Sec. 1.4.4). A non-interacting elementary body does not exist and, therefore, a space with only one geometrical position in the image (picture of this elementary non-interacting body) is not defined; a feature that is not defined cannot be observed. This is in particular valid within our assumption-less observations in everyday life, which take place spontaneously.

Nevertheless, an empty space or a space with only one body is thinkable. We have only to delete the geometrical positions (crosses) in Fig. 10 and an emptiness appears and we come to Fig. 12. In other words, an empty space is "thinkable" and can be defined formally, but it is not "observable".

This in particular means that the space is not permanently installed in the brain. Space and time only appear (in the brain) when there is actually something (objects of the world outside) to picture. What is the reason for that? The answer is straightforward: Each species developed in connection with the laws of evolution, and these laws are dictated by the so-called "principle of usefulness". An observation of an empty space (space-time) would not be useful because it does not contain any information which the observer needs. An empty space is completely useless.

Also the observation of one body, which is alone in the cosmos or does not interact with other bodies, is not possible; the minimal information Eq. (1) does not allow that (see also Fig. 7). The reason is obvious: Such an isolated body is not relevant for an observer because

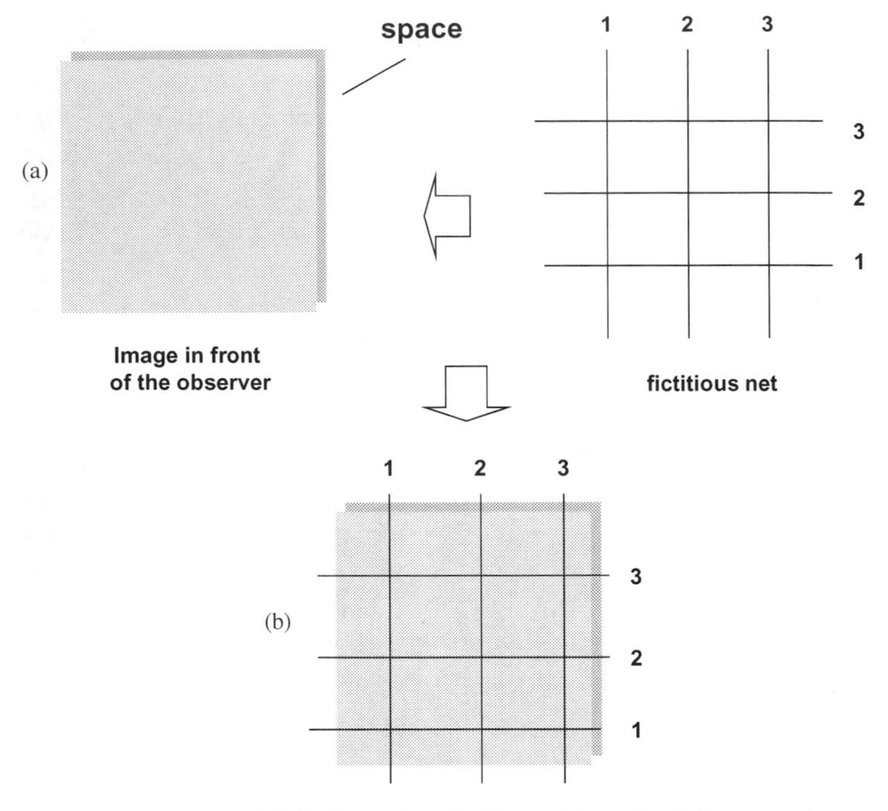

Fig. 12 An empty space is thinkable, but not observable. When we delete in Fig. 10 the geometrical positions (crosses), an empty space appears. This shows that an empty space is conceivable and formally definable. Within our assumption-less observations in everyday life, which take place spontaneously, an empty space never appears in front of us.

it is not able to create physically real effects. This can only be achieved in connection if the body is related to other bodies. Thus, such systems are useless for a human being and possibly for other biological systems too.

The existence of isolated bodies would therefore be against the principles of biological evolution. In other words, a theoretical conception of the world should not allow the existence of isolated bodies. In conventional physics (Newton's theory, Theory of Relativity, quantum theory, etc.) the existence of isolated bodies in space is not forbidden. This is however not the case within projection theory;

here isolated elementary systems do not exist and, if they are more complex with an inner structure, they can exist but are not observable (see in particular [1, 2] and Appendix F).

1.5.4 Space and Time: A Brief Summary of the Facts

The inner world, i.e., the image of the outside world, which appears spontaneously in front of us in everyday life, comes into existence as follows: The observer interacts via his sense organs with the reality outside in order to get the necessary information about it. This information enters the brain of the observer, and the brain constructs an image of reality outside; an example is given in Fig. 1. The construction of the image takes place unconsciously, i.e., without any conscious actions of the observer. The inner world in front of us appears spontaneously.

However, the observer can only interact with entities that are physically real. So, the observer cannot interact with the space-time elements x, y, z at time τ because they cannot be considered as physically real quantities. This in particular means that the elements x, y, z at time τ could not be transferred from the outside world to the inner world.

Space and time cannot exist outside in the form of a container in which real matter is embedded; it is simply not possible to imagine such entities that are not physically real (here space and time) but are, on the other hand, able to gather material bodies, which are the source of physically real effects.

Clearly, our impressions in front of us are configurations in space at a certain time τ (Fig. 1), and these configurations reflect certain facts of the world outside, but — and this is important — the space itself and also the time are inventions of the brain because space and time cannot be considered as physically real quantities. We already mentioned above that we cannot put a piece of space or a piece of time on a table. Nobody is able to do that, and nobody can invent a procedure for that. Such a ridiculous situation is illustrated in Fig. 6.

In connection with space and time, in particular with the space-time elements x, y, z at time τ, our analysis has led us to the following two important features:

1. Physically real entities such as, for example, material bodies of mass m cannot be embedded in space and time because the elements x, y, z and τ are not physically real. From this statement directly follows that the space-time elements x, y, z and τ may not appear in the world outside.
2. If they would nevertheless be also elements of the outside world, the observer could not observe them because they are not physically real. In other words, the space-time elements in the image (the impression in front of us) have to be exclusively considered as the invention of the brain of the observer.

In this way we directly come to "projection theory" because only the inner image can be used for the analysis of the space-time configurations of the world. This is in contrast to the "container principle" where we work directly on the basis of the configurations in the outside world (an example is given in Fig. 3), which we consider as "basic reality". Newton's equations directly refer to this kind of reality, i.e., on configurations like those that are given in Fig. 3.

1.5.5 Statements about the World Outside

The world in front of us, the image of the world outside, is the most direct fact about realty. It is an inner world and reflects certain properties of the world outside. Since space and time do not exist in the world outside, there can be no one-to-one correspondence between the facts in front of us and the world outside; a one-to-one correspondence is only given within the container principle (see in particular Fig. 2 and Fig. 3).

These direct impressions in front of us come into existence spontaneously within assumption-less observations in everyday life, i.e., they come into existence without any intellectual operation. No theoretical model may be in contrast to these spontaneous impressions; they are — in other words — the basis for everything. There is no reason to have doubts concerning the direct configurations in the image. Clearly, delusions cannot be excluded but normally the images are reliable. If such images would not be reliable, humans would have probably died out during the course of biological evolution.

This is an assessment about the inner world. But what can we say about the structure of the outside world? In contrast to our inner-world experience, we have no direct access to the world outside. In order to be able to say something about the reality outside we need theoretical concepts. In this connection a "dialogue with nature" is necessary which is in principle an intellectual process. Let us discuss some details in the next section.

1.6 DIALOGUE WITH NATURE

Essential principles of physics are the reproducibility and objectivity of its findings. These points are believed to be fulfilled in the case of a purely material reality. That is, objects which obey physical laws consist of matter; only material reality is the subject of natural sciences, at least with regard to their main trends. Here the following relevant question arises: How do we arrive at an idea of the physical world outside?

1.6.1 Theoretical Conceptions

There is basically only one possibility: We create a physical conception of the world outside by thinking and this gives rise to questions, i.e., questions that are put to nature itself: That is to say we carry out specific experiments and the deflection of the pointer on the measuring instrument is the answer to our questions. If the theory describes the experiments sufficiently well, we may conclude that the elements of this theoretical conception reflect the structure of the world outside. In this way we map a theoretical conception of the "world outside" onto the "inner world". The principle is schematically represented in Fig. 13.

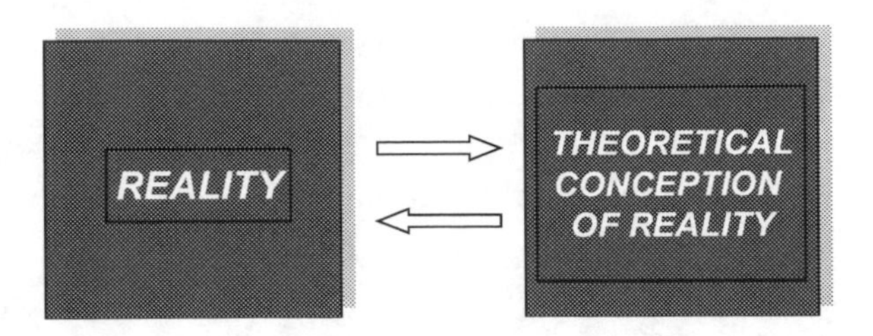

Fig. 13 The inner world is directly in front of us without any intellectual operation of the observer. But what can we say about the structure of the outside world? We have no direct access to the world outside. In order to be able to say something about the reality outside we need theoretical conceptions. A "dialogue with nature" is necessary, which is an intellectual process. What does dialogue with nature mean? Answer: We create a theoretical conception of the world by thinking and we check this conception with the help of measuring instruments.

This is the only way to construct configurations for the world outside. Actual reality outside is in principle not accessible to an observer, and this is because we are not able to take an image-independent point of view, i.e., there is no external point of view which would enable a direct observation of the world outside. We never recognize a theoretical conception spontaneously without thinking.

This is exactly the same procedure how the biological cognition apparatus develops the world in front of us from the information, which we obtain from the outside world through the five senses (Fig. 1). However, we can extend these worlds in front of us, which appear without thinking by consciously developed theoretical conceptions. These conceptions are considered to be physical if we are able to prove them by means of suitable measuring instruments. In this way we obtain — besides our macroscopic impressions — space-time images for atoms, molecules, etc. that are more sophisticated and detailed than our everyday life impressions, although they are not accessible by our five sense organs.

1.6.2 Remark Concerning Metaphysical Elements

What about the structure of a theoretical conception? May such a theoretical conception contain metaphysical elements? In principle yes.

However, then we cannot exclude that these metaphysical elements only appear in the world outside but not in the image in front of us. That is, a "fairy" could be outside but not within the image; we already pointed out that only physically real elements can be transformed from the world outside to the inside world.

There are then no metaphysical elements in the theoretical conception of the world, if with every element of the theory an element-specific reaction of the measuring instrument comes into existence. If that is basically not possible, there is no counterpart of the element of the theory concerned in the material outside world, i.e., such an element is metaphysical in character. Then, we can reject the theoretical conception or we allow that also metaphysical entities are possible within the world outside, but we do not observe them in connection with the world in front of us (Fig. 1), i.e., within the assumption-less observations in everyday life.

If we allow the existence of metaphysical elements within a theory, such a theoretical conception would describe a more general state of the world outside, more general than that which we detect with measuring instruments. We will deepen this point in the forthcoming sections.

1.6.3 Fictitious Realities

Three things are relevant when we try to assess the observer's relation to the world outside. The world in front of us, the image of the reality outside, is the most important fact that the observer can have about the reality outside. This image is a configuration in space and time. The second point is that space and time cannot be entities of the real world outside, and we can say nothing "directly" about this reality, i.e., about true (basic) reality; this world is not accessible to an observer because an image-independent point of view is not possible. We are only able to say something about the reality outside "indirectly" with the help of theoretical conceptions, that is, on the basis of intellectual imaginations. These theoretical conceptions have to be checked with experimental instruments.

We would like to distinguish between three views: 1. Projection principle, 2. Basic situation and 3. Container principle. Let us discuss the main features of these views:

The projection principle

All observers are caught in space and time. Since space and time, i.e., the elements x, y, z and time τ, cannot appear in reality outside, we have to construct other variables when we try to find theoretical conceptions for the reality outside. Let us mark these new variables by the letters R, S, T, Q (Fig. 14). As said, the quantities R, S, T, Q have to be constructed on the basis of the space-time elements x, y, z and time τ because we are caught in space and time. The next step in Fig. 14 (projection principle) is to develop a world view with the help of the variables R, S, T, Q; the results have to be projected on space-time with x, y, z and time τ and we obtain image A. Image A reflects our impressions that we have in front of us in everyday life, but also those facts which have been determined with an experimental setup based on a theoretical concept.

As we already know, the impressions that we have in front of us in everyday life do not appear by the process of "thinking" but they come into existence spontaneously without any intellectual operation. The source for these spontaneous images, which have in Fig. 14 their seat in A, is exclusively the "basic reality" (it is the large full point in Fig. 14 (basic situation), which has to be considered as the "true world".

The reality, which is formed by the variables R, S, T, Q, cannot be considered as basic reality which exists objectively without the variables R, S, T, Q. As we have already outlined above, the variables R, S, T, Q have to be constructed on the basis of space and time with x, y, z and time τ. Since the quantities x, y, z and time τ do not reflect real physical quantities, the variables R, S, T, Q also cannot reflect real physical quantities. Thus, the reality B in Fig. 14 is not a world independent of the observer; let us therefore call it the "fictitious reality". We have to construct these fictitious realities in order

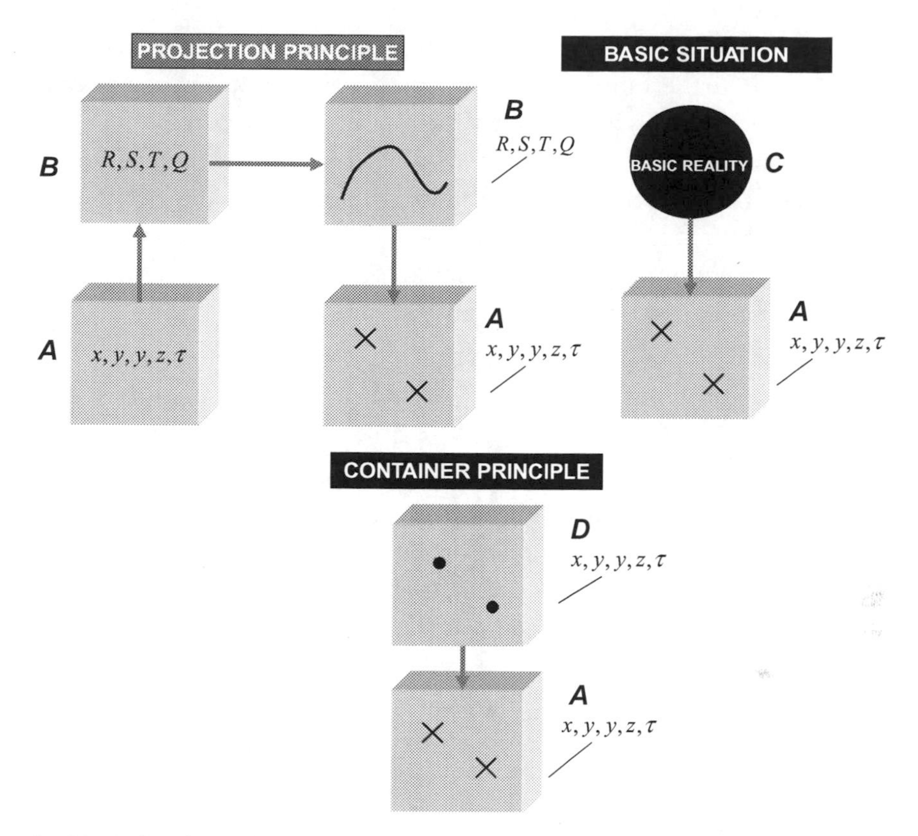

Fig. 14 We have distinguished between three views: projection principle, basic situation and container principle.

Projection principle: All observers are caught in space and time. Since space and time, i.e., the elements x, y, z and time τ, cannot not appear in the reality outside, we have to construct other variables, which are marked here by the letters R, S, T, Q. Because we are caught in space and time, the quantities R, S, T, Q have to be constructed on the basis of the space-time elements x, y, z and time τ. Then we develop a "world view" with the help of the variables R, S, T, Q (see box B), and the results have to be projected on space-time with x, y, z and time τ and we obtain image A.

Basic situation: Basic reality C is projected onto space and time and we obtain a picture of this true reality, which is the seat of all the real physical processes; the observer experiences this image spontaneously without any intellectual operations.

Container principle: In the case of the container principle the structure of basic reality is identical to that of the image; the real masses (full points) are merely replaced by geometrical positions (crosses).

to find realistic images in front of us (this is the contents of box *A* in Fig. 14) and, furthermore, which describe the results of measurements, also described within box *A*. There is no one-to-one correspondence between the fictitious reality (box *B* in Fig. 14) and the image (box *A* in Fig. 14), that is, the image of fictitious reality.

That is all what we can say about the world outside within the frame of projection theory; we never can objectively make statements about the basic, true reality that exists, i.e., independent of human observers.

The basic situation

This view is represented in Fig. 14 (basic situation); the basic reality *C* is projected onto space and time and we obtain a picture of this true reality, which is the seat of the real physical processes. The observer experiences this image spontaneously without any intellectual operations. Within this basic reality the observer, the measuring apparatuses, trees, car and all the other things are positioned as true, objectively existing things; the true physical operations take place in this basic reality.

However, the observer is principally not able to say something about the structure of all these things in the basic reality. He is, in other words, not able to recognize them, because a picture-independent point of view is not possible and even unthinkable. We only recognize images of the basic reality (box *A* in Fig. 14), and we can construct theoretical conceptions on the basis of "fictitious realities".

We can say nothing about the complete contents of basic reality and we also cannot know the transformation laws that transform the information from basic reality onto space and time, leading to the "picture of reality" which we experience spontaneously when we restrict ourselves on assumption-less everyday life observations. This is exactly the difference between the true (basic) reality and fictitious realty. We never observe facts of the true (basic) reality. The reason for this fact is dictated by the principles of biological evolution, which is treated in more detail in the forthcoming sections.

The container principle

In Fig. 14 the point of view within the frame of the container principle is given. We have rejected this principle because the real physical objects cannot be embedded within a container that consists of space and time having the elements x, y, z and time τ, which cannot be considered as physically real quantities. In the case of the container principle the structure of basic reality is identical to that of the image; the real masses (the full points in Fig. 14, container principle) are merely replaced by geometrical positions (the crosses in Fig. 14, container principle). In other words, within framework of the container principle we know the transformation laws that transform the information from basic reality onto the space and time leading to the "picture of reality".

1.6.4 Immanuel Kant

The connections, which we have obtained between the image of reality, the concept of fictitious reality, and the role of true (basic) reality, is undoubtedly close to the position of the philosopher Immanuel Kant. According to Kant, space and time are exclusively features of our brain and the world outside is projected on it, as we worked out in connection with the projection theory. Then also, Kant's ideas lead to the fact that the material objects that occupy space and time can only be geometrical pictures. Unfortunately, Kant was not able to give mathematical formulations for his basic thoughts, i.e., for the relationship between space-time and the real world. This is obviously the reason why most physicists ignore Kant's philosophy. In Appendix D more facts are given.

1.7 SCIENTIFIC REALISM

Although we have rejected the container principle, let us deepen the situation given by this principle, i.e., in connection with the situation given in Fig. 14 (container principle). The container principle suggests that the structures and characteristics in the picture (box A in Fig. 14) are identical with those of the world outside. That this feature must be a fallacy has been pointed out in the sections above.

In fact, most people assume automatically that the things in front of them is the material reality itself, and those which are conscious of the fact that it is only a picture normally assume, as a matter of course, that the structures and the other characteristics in the picture are identical with the material reality outside. It is evident that there are no material objects in the picture (box A of Fig. 14) but only geometrical structures, and that what we see in front of us is a picture of reality and not the material reality itself. However, as already mentioned, it is assumed by most people that the structures in the picture are identical with those in the actual reality outside.

Which kind of truth is ultimately described by theoretical physics, in particular by Newton's mechanics? Is physics actually in the position to describe that what we have called the "absolute truth"? Let us explain this point by means of Newton's equations of motion.

1.7.1 Newton's Equations of Motion

We want to study the situation by means of an example, namely the earth and its motion around the sun. Earth and sun interact, i.e., they attract each other, and for this attraction Newton could formulate a mathematical relation. Furthermore, if we know the position and the velocity of the earth at a certain but arbitrary time, then the movement of the earth around the sun is determined for all times τ, at least in principle.

One knows, in other words, at which position in space the earth will be in the year 2020 or where it was, for instance, in the year 1900. In order to calculate these data, we only have to solve the respective equation, which is quite simple in its structure and has the following form:

$$m_E(d^2\mathbf{r}_\mathrm{E}/d\tau^2) = -Gm_Em_S(\mathbf{r}_E - \mathbf{r}_S)/|\mathbf{r}_E - \mathbf{r}_S|^3,$$

where $\mathbf{r}_E$ and $\mathbf{r}_S$ are the positions vectors of earth and sun, m_E and m_S are their masses, and G is the gravitational constant. More elements are not involved in the equation of motion. All the other details in connection with this equation are not of interest for our discussion. Once again, Newton's equation of motion describes the path of a celestial object under the influence of another celestial object, here the path of the earth within the gravitation field of the sun.

Let us assume that the above equation describes the absolute reality. Then, we have to consider the masses m_E, m_S and the gravitational forces between them as really existing in the reality outside. To imagine that, does not mean any problem for the present. Also the solutions of Newton's equation of motion (it is a differential equation) have to be considered as really existing, that is, the possible paths of the mass reflect the deepest ground of reality. This assumption too makes no problems since we actually observe the movements of the celestial objects; we have them directly in front of us. Therefore, we can assume for the present that these paths reflect an absolute fact.

Up to this point one can pre-suppose, that the equations of motion with all their elements and solutions reflect the structure of absolute reality and, therefore, actually describe what we have called the "absolute truth". In other words, reality at the deepest ground is structured as it is formulated by physics.

But we have to be careful, because such kind of scientific realism would consequently mean that the celestial objects would continuously solve differential equations in their movements through space, since Newton's equations of motion are differential equations, as already mentioned above.

Such an idea must however be considered as absurd. Where is the computer hidden, which would do this work? Who does this work with lightning speed and simultaneously everywhere? By the way, the mechanism with which nature solves the differential equations, would have to be delivered by the theory itself. In other words, it should be the content of the theory itself. This is however not the case.

Such a scientific scenario (realism), that the masses incessantly solve in their motion through space differential equations, has to be excluded and must be judged as ridiculous, which is also reflected in the following remark [9]: *As Herschel ruminated long ago, particles moving in mutual gravitational interaction are, as we human investigators see it forever solving differential equations which, if written out in full, might circle the earth.*

This brief analysis makes clear that we, with Newton's ideas, by no means affect what we have called "absolute reality" and "absolute truth", respectively. That is true for all theoretical developments in physics, even for the newer theoretical developments. This is because everything in natural science is formulated in accordance with Newton's basic way of thinking, which is characteristic that the scientific laws are formed on the basis of what is directly in front of the observer. Nicholas Rescher, a distinguished epistemologist, expressed this fact by [9]: *Scientific realism is the doctrine that science describes the real world: that the world actually is as science takes it to be, and that its furnishings are as science envisages them to be .... It is quite clear that it is not ...*

The facts, i.e., the paths and the physical elements together with the effects tied with them, which are described by the above introduced equation

$$m_E(d^2\mathbf{r}_E/d\tau^2) = -Gm_E m_S(\mathbf{r}_E - \mathbf{r}_S)/|\mathbf{r}_E - \mathbf{r}_S|^3$$

only represent a certain kind of truth but cannot be considered as absolute in character. If everything, which is described by this equation of motion, i.e., the paths of the celestial objects, which we have in exactly this form directly in front of us, cannot be the last, absolute truth, then also all the other things, which are in front of us and which we feel by our senses cannot be the last, absolute truth too. This, of course, is also true for that what we perceive of ourselves and other creatures. This result is of fundamental meaning.

In a nutshell, the equations of motion cannot be the last, absolute truth. Since the derivation of these equations have been directly based on that which we have in everyday life in front of us, then these optical impressions cannot be part of an absolute (basic) reality. Thus, a human observer is not able to recognize the absolute truth, that is, the deepest ground of reality. The situation is summarized in Fig. 15 and Fig. 16.

Fig. 15 Newton's equations of motion can be no instrument for the description of the absolute (basic) truth. In particular, we have ascertained that it is hardly realistic to assume that celestial bodies solve differential equations during their motion.

What kind of reality could be involved when we base our descriptions on the equations of motion? Do the impressions that we have directly in front of us actually reflect a certain kind of reality at all or is it "only" a symbolic picture of it?

When we base our considerations on the usual assumption that everything is embedded in space and time, then we may assume that the impressions in front of us directly reflect a real situation with material objects. However, we found out in the preceding sections that this should not be the case. It turned out that it is more realistic to assume that our direct optical impressions are "pictures of reality" but not reality itself, i.e., the objects in space and time are geometrical figures and are not material objects.

Fig. 16 As in the case of Newton's equations of motion, we cannot give any statement about all the other things as, for example a house or a human being. How these things are structured in true (basic) reality remains principally hidden to a human observer.

Remark

Do the theoretical laws of physics, as the equations of motion, describe the real world? We answered this question in the negative. "The theory of science" came to exactly the same result. We already remarked above that Nicholas Rescher came to the conclusion that the structure of the world is not as science takes it to be. What arguments are used in the theory of science? Since this scientific direction is important, we have summarized some of the relevant in Appendix E, in particular those points which underline that the basic reality (the absolute truth) cannot be grasped by science.

Nevertheless, let us already mention here the following: After the so-called "asymptotic convergentism" science progresses by approaching the truth successively: The "final answer" and the "final view" of things, respectively, is gradually approached by the way of an asymptotic approximation. At least until one generation ago, the opinion was firmly established that science is cumulative and the advocates of the scientific method had understood scientific progress to have this cumulative nature. Within this concept the "absolute truth" is set equal to "our ultimate truth". But the asymptotic convergentism cannot be upheld because the progress of science speaks another language.

Essentially two serious objections can be quoted against these asymptotic convergentism:

1. There is no metric to measure the "distance" between bodies of knowledge (theories).
2. There is in general a fundamental change in perspective when we replace a theory by an improved one.

Within the asymptotic convergentism it can rather be assumed that the world is actually as science envisages it to be. Here, fundamentally new pictures of the world are not generated again and again. Here we have one frame, which is filled successively until we have the final, absolute truth (identical with the structures of basic reality). In this case the concept that theoretical terms like electrons, quarks, etc. are actually existing entities in basic reality is justified.

However, as already mentioned, the asymptotic convergentism can no longer be upheld. Instead, science progresses by a sequence of incommensurable schemes of thinking (pictures). In this connection the following point is relevant: Because a metric for the measurement of the "distance" between bodies of knowledge is not definable, there is no possibility to express certain peculiarities of two theories for the absolute truth and for the basic (absolute) reality, respectively. In other words, the results of the theory of science are in accord with the basic statements of the projection theory.

1.8 REAL AND METAPHYSICAL ELEMENTS

The classical equation of motion does not reflect the absolute truth. In the last section we analyzed the situation in connection with Newton's mechanics and we came to the result that it would be too naïve to give the equations of motion this outstanding rank. Since all the newer theoretical conceptions have been developed after Newton's principle, we have to assume that this statement has to be valid also for the other theories, i.e., also these theories should not reflect that what we have called the "absolute truth".

1.8.1 Basic Reality is Metaphysical in Character

The following question arises: How can we expect to find a "world equation" when true reality is not accessible? We have to be careful with our demands. If the absolute truth can principally not be grasped, it is principally not possible to find that what is often called "world equation".

It is generally believed that the absolute truth within Newton's theory is given by the container principle, i.e., the real masses (real bodies) are embedded in space and time and we come to the scheme Fig. 14 (container principle). However, the analysis of Newton's equation of motion led unequivocally to the conclusion that these equations should not reflect the "final, absolute truth". This statement can be generalized and we may claim that no theoretical construction for the world outside can tell us something about the absolute truth. In other words, the absolute truth remains hidden and we come to the scheme in Fig. 14 (basic situation); this scheme with the full large point indicates that the basic reality is principally not accessible to human observers. Therefore, we have to seek recourse to the "projection principle" (Fig. 14, projection principle) with its "fictitious realities". In other words, the container principle is replaced by the projection principle.

The fact that basic reality cannot be observed means that the notion of "basic reality" is metaphysical in character. Within the materialistic point of view such metaphysical systems and elements, respectively, are strictly forbidden. However, within conventional physics, where the container principle is applied, metaphysical elements enter the theoretical description and are unavoidable. This point is important and will be discussed in more details in the next section.

1.8.2 Metaphysical Elements in Conventional Physics

When we work within the framework of the projection principle, which has to be considered as more realistic than the container principle, we leave conventional physics. But let us still stay for a while within conventional physics where the container principle is valid. What is the structure of the world outside within conventional physics? This question has been answered above as follows: We construct a theoretical conception of the world outside and compare its results with certain experimental findings. In the case of good agreement we define the structure of the world outside via the elements of the theoretical conception. This is an "indirect" procedure; a "direct" observation of the world outside is not possible because a "picture-independent point of view" is not possible.

Within the materialistic point of view only real matter is positioned in the container outside, consisting of space and time. (In modern conventional physics the container is also filled with real fields). Most physicists and also other people are strict materialists, and this is because only what is considered to be real can be felt. Only those bodies in front of a human being, which are observable with the five senses or with measuring instruments, are considered to be real. This in particular means that a materialist is firmly convinced that just these observation procedures deliver the complete information about the world outside.

Once again, what is the structure of the world outside within the container principle of conventional physics? Answer: We construct a reliable theoretical conception of the world outside and we define the structure of the world outside through the elements of this theoretical conception. Then, "all" the elements in the formulas of the theoretical construction are positioned within the world outside, i.e., within space and time.

But what can we say about the elements in the formulas as, for example, those within Newton's mechanics? What is the nature of these elements? Are they compatible with the materialistic point of view? They are not because some of the elements are metaphysical in character, i.e., they cannot be observed with our five senses and also not with suitable measuring instruments.

Let us explain this point by Newton's mechanics, although all the other theories of conventional physics are equally concerned and, therefore, we may state that metaphysical elements are unavoidable in conventional physics.

As we have already outlined above, within Newton's mechanics equations like

$$m_E(d^2\mathbf{r}_{\mathrm{E}}/d\tau^2) = -Gm_E m_S(\mathbf{r}_E - \mathbf{r}_S)/|\mathbf{r}_E - \mathbf{r}_S|^3$$

are relevant. In this equation $\mathbf{r}_E$ and $\mathbf{r}_S$ are the position vectors of earth and sun, m_E and m_S are their masses, and G is the gravitational constant. More quantities are not involved in the equation of motion. The elements of this formula dictate the contents of the world outside. i.e., they dictate the elements that are embedded in space and time. We have no other possibility.

Besides m_E, m_S and the position vectors $\mathbf{r}_E$ and $\mathbf{r}_S$, the interaction between m_E and m_S appears. In other words, the formula says something about the interaction mechanism between the two bodies. According to Newton's mechanics it appears to be the case that the interaction between the masses m_E and m_S acts across space instantly. This is because in the gravitational law there is a relationship between the spatially separated positions of the masses and no intermediate position appears in the law. This suggests the view that the interaction works at a distance, i.e., the forces come about through an "action-at-a-distance" effect as follows: The gravitational force reside in a body (for example the sun), but comes into effect at the location of the other body (for example the earth). The space between the bodies in this view is free of gravitation.

It is important to underline here that this concept of interaction, which is expressed by an "action-at-a-distance", cannot be checked, i.e., these real effects cannot be observed by measuring instruments and not with our five senses. In a nutshell, actions-at-a-distance are processes that are metaphysical in character.

No doubt, action-at-a-distance is an element of the mind. But it is also an effect in reality, when we work within the frame of the container principle. That this effect does actually take place in reality, must however be supposed and cannot be checked experimentally. In other words, it is a matter of belief.

We recognize that the notion "metaphysical element" comes not first into existence in physics through what we have called the "basic reality" in Fig. 14 (basic situation). The notion "metaphysical element", as it is used in natural science, has its origin in conventional physics and comes not first into play through the projection theory where we have to distinguish between the fictitious reality and the basic reality (Fig. 14).

In order to be able to find a classification scheme for the arrangement of physically real quantities and metaphysical elements we have to investigate the role of metaphysical elements and metaphysical statements in more detail. Let us do that in the following sections.

1.8.3 Two Types of Statements in Conventional Physics

If we can assign to every element in our theoretical conception of the world an element-specific movement of the pointer of a measuring instrument, then we can with good regard treat the constructed theoretical conception of the world as free of metaphysical additions. If that would be possible, the materialistic world view has to be accepted. But we cannot, as we will recognize in the following.

What does measurement in general terms mean? Let us take Gottfried Falk's (1922–1990) comment for answering this question [10]: ... *Measurement means nothing else than to express an observation in numbers. To this extent physics can be regarded as an effort to record nature — the world — using the concept of number.*

Can what we call physics, however, actually be defined in this way? Certainly, physics uses a large number of formulae, that is, (formalized) statements, which operate with the concept of number — for example, the law of gravity, the general gas equation, the Balmer formula of hydrogen and many more — but there are also physical statements in which number plays a subordinate role (and something not the slightest role at all), such as the assertion that all matter is composed of specific elementary building blocks (elementary particles) or that it is impossible to carry out processes in the world which violate the two principal laws of thermodynamics. These assertions which are not immediately recognizably tied to numbers, and yet, like those just mentioned, they enjoy the reputation of being particularly fundamental.

In this statement Falk demonstrates that there are relevant elements and statements in conventional physics, which are principally not observable. Such statements have to be considered as metaphysical in character. We find further in [10]:

Logically, a fundamental difference exists between these two kind of physical statements. It is based on the fact that all observations (to which measurements also belong), as also the passing on of observations, can in principle be done by means of a finite number of data. The infinite, in whatever form it appears, is never a result of observation, but is always a free addition by us in describing the observed phenomena. Statements in physics about nature are therefore in principle of two fundamentally different types: Those which work with a finite set of data, and those which require in quantitative formulation infinite amounts of data. I call the first scientific, the second transcendental or metaphysical because they claim facts, which exceed in principle our experience, since they require more data than our observations are ever able to deliver.

What is surprising now is that such trusted and highly regarded statements as the two principal laws of thermodynamics are not scientific but metaphysical and this is also the case with the assertion that matter is composed once and for all of absolutely given "final building blocks", i.e., elementary particles.

If our "conception of the world" is based on material bodies consisting of "final building blocks", i.e., elementary particles, our theoretical world conception is therefore an essentially metaphysical one. Exactly the same is true when we base our description on similar units like strings etc.

Measured with an adequate logical yardstick, the concept of the elementary particle must be classified as metaphysical. One connects the introduction of elementary particle with the idea ... *that matter is composed one and for all of absolutely given "final building blocks", i.e., elementary particles* [10]. However, this statement exceeds in principle our experience, that is, it is definitely metaphysical in character.

Then, we may state the following: When we introduce what we have called above "basic reality", which has to be considered from the start as a metaphysical system, no qualitatively new feature enters natural science since conventional physics is full of metaphysical elements and statements, as we have just recognized in connection with Gottfried Falk's analysis. However, the metaphysical "objects" that appear in conventional physics should not be considered as a drawback. They belong to the main essence of science, and we obviously cannot work without them.

Just the notion of "elementary particle" underlines that everything in physics is in motion up to the most basic ideas. In Werner Heisenberg's (1901–1976) opinion, it is quite arbitrary to distinguish between elementary particles and composite systems. Concerning this point we find the following comment in [11]:

Particles can be created and annihilated, they can be converted into other particles, and so the distinction between elementary particles and composite systems is lost. It is important to make clear to oneself, that the entire basis for an atomic materialism, in the philosophical tradition of Democritus, is thereby destroyed. The concept of "separation" already loses its meaning if a very small piece of matter is to be divided; for the dividing process requires energy, and if the energy is sufficiently large, the result of the process can be interpreted rather [...] production of new particles rather than a division of the original piece of matter.

In other words, Heisenberg was sceptical concerning the concept of elementary particles. However, as we know, the elementary particle concept, conforming to a building block principle, is very successful and one is reluctant to reject it so simply. One probably should not treat it with an absolute certainty (*... one and for all from absolutely given "final" building blocks* [10]). Within a certain framework one can certainly assume that matter is composed of "little units" which one could call elementary particles. But these little units must certainly not be identified with "final building blocks".

The concept of the conventional elementary particle physics is based on the "container principle" (Fig. 14), as with all the developments in conventional physics. But — as we know — this container principle has to be disqualified as unrealistic. We have outlined above why that is the case. The "projection principle" obviously fulfils the basic features of space and time more realistically then the "container principle". We therefore expect that within the projection theory new aspects appear in connection with the notion of the "elementary particle". In fact, a detailed analysis revealed that independent elementary systems may not exist within the projection theory [1, 2].

1.8.4 The Infinite

The statements by Gottfried Falk made clear that the infinite is of particular relevance in physics. Physics is full of metaphysical additions, not only the concept of elementary particles belong to these additions, but particularly the "infinite".

The infinite can never be the result of an observation, in whatever form it appears. The infinite is not only quantitatively inaccessible but it can also not be experienced qualitatively by the imagination. The infinite arises from pure thinking, whatever thinking may mean — discovery and/or invention. In other words, from the information about the outside world, which the five senses carry to the brain, one obviously cannot conclude the existence of the infinite. The well-known paradoxes of the infinite always appear then when one tries to comprehend the infinite with contemplation (more details are given in [10]). The electrical and gravitational fields are typical examples of quantities with an infinite set of values. According to Falk such quantities (fields) constitute metaphysical elements, since — as we have outlined — only those quantities with a finite set of values can be recorded with established measuring methods.

One could take the point of view that, if a sufficiently large amount (say L, different from infinity) of a physical entity with an infinite set of values has been confirmed by measuring, the remainder will certainly also agree with the theory; that is, the larger L is, the more precisely the physical entity is determined.

This argumentation is however deceptive because it does not consider the particular nature of the infinite. In connection with our discussion a certain feature is of particular interest: A set of infinite numbers is independent of the set L, i.e., there is always a remainder left over, which can contradict the theory, and this remainder continues to be infinite. This is because the following relation is valid: $\infty - L = \infty$. (In other words, the subtraction of a finite number L from infinity is again infinity.) Therefore, if one subtracts some numbers from the infinite set, an infinite set is left, nonetheless. The infinite *is an inexhaustible jar, a miraculous jar recalling the miracle of the loaves and the fishes in Matthew 15:35* [12].

All this is mystical, mysterious, metaphysical. *The infinite is that which is without end. It is eternal, the immortal, the self-renewable, the apeiron of the Greeks, the Ein Sof of the Kaballah, the cosmic eye of the mystics which observes us from the godhead* [12]. How can a closed physical theoretical conception be formulated without the infinite, without this miraculous jar? Probably never. *We have infinities and infinities upon infinities; infinities galore, infinities beyond the dreams of conceptual avarice.* [12]

Consequences

Conventional physics is considered to be a basic science, and this means that all the other phenomena in natural science, for example in chemistry and biology, derive from it. An essential question arises: Is there anything at all which is free of transcendental peculiarities and metaphysics, respectively? If we measure with the logical yardstick previously discussed it is definitely not. However, we already mentioned above that the metaphysical "objects", which appear in conventional physics, should not be considered as a drawback. They belong to the main essence of science. Materialism and positivism (see in particular Sec. 1.9) deny that.

1.8.5 Further Remarks on Materialism

The roots of the view, or rather endeavour, to ascribe to all phenomena in the cosmos a material basis, is essentially due to our habit of understanding anything we perceive as a body, or to consider the cosmos as a collection of bodies, which are set in space. In this connection body and matter are considered de facto as the same. In the context of this view the question of how matter is constructed becomes one of the most fundamental issues. However, this point seems to be overestimated because we are obviously not be able to construct a theoretical world view without metaphysical entities. Therefore, the metaphysical entities and their role in science have to be investigated as well.

The strict materialistic world view does not allow the appearance of metaphysical elements. Since a statement about the composition of the world outside has to be based on a theoretical conception — just this conception determines the composition of the world outside. As we have discussed, these theoretical views can obviously not be constructed without metaphysical entities. From this point of view, not only do the material objects belong to the world outside but also certain metaphysical quantities, defined through the theoretical conception.

As we have established above, conventional physics is full of metaphysical elements, i.e., it contains "objects", which in principle cannot be experienced, at least not with conventional observation methodology, which is exclusively based on the five senses of the observer and on measuring instruments. This is an unsatisfactory situation in the context of the conventional norms.

Fact is that this observation methodology is considered to be indisputable which in turn has led to the view that those "objects" and elements that cannot be experienced by this tightly anchored perception system do not exist and must therefore be rejected.

The demand of the materialists and the followers of positivism was thus to eliminate all metaphysical entities from physics and the natural sciences, since they are not accessible by measurements and to perception by the five senses.

But is it necessary at all to make conventional observation methodology an absolute criterion? If not, on what can we base our considerations? Once one leaves the realm of conventional (materialistic) observation methodology, one very quickly ends up — in the general opinion — in a bottomless pit. However, the limitlessness which metaphysical elements may bring to bear on the situation is no reason to demand that such elements have to be regarded as non-existent. Conventional observation methodology offers safety and certainty, and there is obviously a psychological need for safety and certainty. The existence of metaphysical entities would disturb this kind of order, and this is because they are non-observable and therefore the entire scenario becomes uncertain.

As already mentioned several times, observations are carried out through the five senses and with the help of measuring instruments. This procedure ensures reproducibility and objectivity. No doubt, these are important features.

Whereas an assumption-less observation is possible with the five senses, the use of measuring instruments already presupposes an intellectual or theoretical idea of what one wants to measure. The experiment has to give an answer to questions the observer has formulated.

In this connection it is important to note that there is practically no basic idea (theoretical conception) in modern physics, which would have evolved without the use of metaphysical elements, i.e., without scientific ideas like the infinite, the concept of elementary particles, etc. The point is that we actually work with these non-observable entities, even if only some of the theoretical conceptions are objectively observable with the established observation methods.

Let us summarize the situation: On the one hand, a conception or theory can often allow one to make very informative and even exact measurements. On the other hand, the theoretical conception contains metaphysical elements, which are not accepted as components of the real world, as the success of the scientific method is essentially based on the idea of eliminating metaphysical entities. This is obviously impossible.

Physicists seem to get along quite well with this inconsistent stance. They simply hesitate to reject theories containing metaphysical entities, since in many cases results of those theories are in a large measure actually objectively observable with established observation methods.

However, a positivist would not accept such a situation, and this must be recognized. Positivism is often considered as an attractive standpoint even for those scientists who have somehow accepted a "mixed" reality consisting of observable and metaphysical elements. However, we will recognize below that both positions are unsatisfactory.

A person who only accepts a material reality without metaphysical additions, obviously lives in a world that is incomplete, but this world offers him security and certainty. In this context, a relevant question is essential: Can security and certainty be bought with incompleteness? Concentration on the five-senses-world is certainly also connected with the fact that such a world is easy to deal with. As already stated above, security and certainty are basic psychological needs of mankind. But not only this, the materialistic level guarantees exercise and power and also self-portrayal; all these things require situations which are largely free of doubts and ambiguities. No doubt, such situations are disturbed through the existence of metaphysical additions.

Security and certainty, which the material level seems to offer, is possibly bought by accepting incompleteness. This is however not felt by the observer, because material reality is dominant over the mental level. Furthermore, the material level was developed first during the evolutionary processes in nature and therefore must have a certain completeness in itself. All these points will be considered in more detail below.

1.8.6 Concluding Remarks

A theoretical conception of the world should then be free of metaphysical entities, if an element-specific value on the measuring instrument can be assigned to each element of the theory; then one can equate the statements of the theory with events in the world outside. In this connection two statements are of particular interest:

1. To measure means to express an observation in numbers [10]. More precisely, with the help of a finite set of numbers; the infinite cannot be experienced. That is, the infinite is a metaphysical entity.
2. In addition to this point there are statements in physics in which numbers only play a secondary role as, for example, the concept of elementary particles. According to Falk, such statements also belong to the category of metaphysical elements, because they are not accessible to measurements. That is, they cannot be expressed with the concept of number.

From this follows quite clearly that we cannot assign to every element of the theoretical world view an element-specific measurement. One reason is that within present-day physics a lot of entities appear which need an infinite set of values. For example, in conventional physics the concept of field is important, and each field consists of an infinite number of elements. Apart from that, there are a lot of assertions in physics, which should be classified as belonging to point 2.

We have to assume that we will never have a theoretical conception without metaphysical entities and metaphysical statements. This must not be a problem. However, we need a new "classification scheme" for the understanding of all these things; in particular for the purpose to overcome the unsatisfactory situation in connection with mixtures of material and metaphysical entities. Such a new "classification scheme" will be proposed and discussed below on the basis of the "projection principle" which seems to be more realistic than the container principle.

Conventional methodology recognizes observations on the basis of the five senses and at the level of measurement as valid. This procedure guarantees reproducibility. We have however noted above that there is practically no theoretical conception in modern physics which could be constructed without metaphysical entities. This is the case even if some effects of a theoretical conception are objectively observable with the established observation methods.

Contemporary physics therefore goes beyond the objectivity which it claims to represent. From its point of view the mixed states, consisting of observable and metaphysical elements, must be considered as unsatisfactory. Two contrasting kinds of stand can be taken on this issue:

1. Conceptions of the world which contain observable as well as metaphysical entities yield unsatisfactory theories (point of view of materialists).
2. One can however also take the view that such conceptions of the world describe more general states, more general than those recorded by measuring instruments; according to this view the world can only be described incompletely by measuring instruments and the five senses alone.

The incompleteness of the world we observe with our five senses (Fig. 1) is obvious and leads to naive and deceptive ideas of matter. The material world itself, as we imagine it directly, has its roots in the observations performed by the five senses in everyday life, i.e., at the level of direct and unprejudiced (assumption-less) observation.

At this level we regard bodies and matter as de facto the same [10]. Bodies, i.e., matter, as they appear directly to us, present themselves as a continuous medium, and their boundaries in space that determine their form.

"Continuous medium" here means densely packed matter, as it appears in all unprejudiced, assumption-less observations in everyday life. However, it turned out that this idea had proven to be deceptive, because what appears as densely packed has in fact an atomic structure, that is, it is in fact empty space with an insignificant fraction of real matter. In other words, what we have directly in front of us in everyday life within assumption-less observations (Fig. 1) cannot be considered as a complete representation of the world.

Positivism is a philosophical direction which ignores to a large extent what is beyond the five-sense experience. Since positivism is still an often discussed point of view, we would like to give some remarks on this specific philosophical direction.

1.9 POSITIVISM

Positivism is a branch of philosophy which starts from the "positive", whereby the term "positive" is understood as the certain and unquestionable. The positivist essentially limits his research to these points; metaphysical debates are from this point of view theoretically impossible and practically useless. The strict materialism is close to positivism.

The basis for all knowledge for the positivist is therefore certainty and doubtlessness, whereby he believes that this can only be experienced by his senses; he therefore only learns something about the world by opening his eyes and ears. In his view he is born without knowledge and all knowledge about the world is exclusively realized through the five senses. In other words, the positivist does not believe that there is more than his sensory impressions. No doubt, positivism is a somewhat strange kind of philosophy and has been declined and criticized very often.

1.9.1 Solipsism

The positivist is basically a solipsist, i.e., he is a person who only recognizes the content of his own consciousness as existing. Karl Popper outlines in [13] that all people who are not solipsists, but are nevertheless positivists, have already made a compromise if they concede that other persons exist in addition to themselves.

Bertrand Russell (1872–1970) went a step further with his "solipsism at the moment", for he does not even admit that there is a past and a future. Russell does not even trust the memory of past experience, nor the projection of the experience into the future, i.e., for him these elements are not sufficiently sure and reliable.

These ways of thinking, which form the basis of positivism, are certainly consistent in their logic, even including the "solipsism at the moment", as long as one recognizes that certainty and the unquestionable actually exist.

1.9.2 Karl Popper

Karl Popper (1902–1994) says very aptly [13] that one must understand positivism as a far-reaching generalization of the concept of induction, i.e., the conclusion of the "general" from the "particular". In [13] we find the following comment:

Positivism is actually the view that the idea of going from the particular to the general must be applied so strictly that we start from our direct observations, especially from our elementary feelings produced by the senses. And from these experiences we then gradually develop our knowledge of the world and our theories. That is positivism. Mach developed these conceptions very strongly, and this is particularly expressed in his famous book "Die Analyse der Empfindungen". Feelings are for him the elementary observational experiences, and in his early phase Einstein was also a follower of this view — although later he reacted very strongly against it, and then arrived at a completely different view of the world.

Positivism is thus essentially based on the concept of induction, i.e., it is concluded from the particular to the general. However, experience tends to speak against the concept of induction, and in fact militantly so, that such a view can hardly be maintained. Too many breakthroughs in research have been achieved by deduction, that is, the particular has been inferred from the general. For example, it is accepted that Kepler's laws can be derived from Newton's mechanics by deduction. Here are just a few more examples.

1. In connection with Albert Einstein, "mathematical intuition" was a relevant point and the starting point of his theoretical investigations; he was guided by the mathematical beauty of a theory. Using this theoretical approach he then deduced the various observable facts and compared them with experimental results. This should be considered as the only scientific and acceptable way in science.
2. The discovery of the DNA in 1953, certainly one of the greatest discoveries in the last century, was also made in a deductive way: Watson, Crick and Wilkins designed the macro-molecule step by step in advance and proved it experimentally afterwards.
3. In accordance with Kepler's first law the planets move on trajectories with a form that is described by ellipses. The ellipse was a fundamentally new idea as a solution for the problem of orbital paths; it does not follow from the circle by induction. The orbit can be a circle if we assume that the elements of the world can only be experienced without exception through the senses, which is precisely the position of the positivists. Then, the intuitive basis for the movement of the planets is the movement of a wheel. Kepler was convinced of the harmony of the world. This conviction (and not the wheel) possibly helped him to objectify the ellipse as the path of orbit.

Experience tends to speak against the concept of induction. Such a view can hardly be maintained. However, for a lot of scientists the concept of induction is still convincing, and the reason for that is probably the belief that the facts can possibly be re-interpreted some day in favour of the concept of induction.

1.9.3 Groping in all Directions

In an interesting discussion with Franz Kreuzer, Karl Popper underlines significantly that the concept of induction has to considered as wrong [13]. In [13] an instructive statement is quoted:

Groping in all directions. I do not favour the picture of science as gathering observations and distilling the laws from them in the same way, as Bacon said as wine is pressed from grapes. In this conception the grapes are the observations and the wine obtained from them is the generalization, the theory. This conception is utterly wrong. It mechanizes the creative act of human thinking and inventing. To make that point clear is most important to me. Science proceeds in a different way, that is, it proceeds in such a way as to test ideas and world views. Science derives from myths. This can be seen very clearly in connection with the early scientists, i.e., the early Greek, pre-Socratic philosophers, who were still strongly influenced by myths.

In other words, not the concept of induction (the general is concluded from the particular) should be correct but just the opposite, i.e., the particular is concluded from the general (concept of deduction).

"Groping in all direction" means testing a physical view of the world via a dialogue with nature (Sec. 1.6) on a certain level of reality. In this way a world view can be improved and, if necessary, also be rejected. The theory should be reflected in many and, as far as possible, in all physical situations, and that has to be verified. Such a process can be called "objectivation by thinking".

A theory that is useful should be constantly reflected in many variations of thinking and different experimental configurations. This should also be valid for other fields. In [2] we remarked the following: "*Similar criteria — perhaps not so severe — should be fulfilled in connection with the products of imagination (fantasy). In this connection we outlined above: A literary picture very often also sums up reality in a single "image" (metaphor). This however can be applied in many situations of life, and the same "image" (metaphor) is reflected in the experience of many people. We may therefore state that also the author of a narration can find a single "image" by a process of objectivation, just as in the case of a physical theory.*"

1.9.4 Concluding Remarks

When we apply this direction of thinking on our consideration we come to an essential conclusion: The "conceptions of the world" cannot be deduced from the space-time images that we have directly in front of us in everyday life (Fig. 1), and this is because the conceptions of the world are more general than these spontaneous space-time images.

1.10 CLASSIFICATION WITHIN CONTAINER THEORY

Positivism and materialism would not have a real basis without the container principle. Once again, within the frame of this principle the impressions in front of us (for example the specific image of the world in Fig. 1) are identical within the world outside; the geometrical positions in the image are merely replaced by the real material bodies. That is, the real world is embedded in space and time. This view has been clearly rejected in the discussions above. The details given in Sec. 1.7 support that; in particular, in this connection the statements by Karl Popper are instructive.

Nevertheless, in order to find a convincing "classification scheme" for the material and metaphysical entities that appear in the theoretical conceptions of the world, let us still investigate some basic questions with respect to the container principle. Just these theoretical world-conceptions are of basic interest and relevance when we try to find the configurations of the world outside. However, the interpretation and also classification of such theoretical world views become more than problematic when we work within the frame of the container principle introduced above, i.e., when we pack up the physically real world within space and time.

1.10.1 Where are the Metaphysical Elements Positioned?

Within the container principle not only the real masses occupy space and time, but everything else which is important for the organization of this matter. All those "things" should somehow occupy space and time that are involved in the construction of the theoretical concept which determines the features of the world outside. These things must have a place in space and time. This is however not possible. The reason is relatively simple to understand.

These theoretical conceptions contain metaphysical entities like the infinite and, furthermore, there are also physical statements such as the assertion that all matter is composed of final elementary building blocks (elementary particles) or the statement that only processes are possible in the world, which fulfil the two principal laws of thermodynamics.

Such metaphysical elements and basic statements are not only in the "mind of the observer" but belong also to the "world outside". Within the frame of the container principle they should also exist without the observer. In other words, within this view the "conceptions of the world" exist twice: Inside the observer's brain and also in the world outside, i.e., within the container made of space and time.

Example

Within Special Theory of Relativity the mass m varies with velocity, and the well-known law

$$m = \frac{m_0}{\sqrt{1 - v^2/c^2}} \tag{5}$$

describes this effect; here m_0 is the rest mass, v the velocity and c is the velocity of light. The law Eq. (5) is not only in the brain of the observer but in the world outside as well. The real world outside is organized in accordance with the physical laws and Eq. (5) belongs to these laws. Since the observer does not tell the real physical entities outside how they have to behave, the physical laws must exist also outside in the container independent of the observer. The laws are somehow within the space outside.

Clearly, not only those parts of a law exists outside, which are actually realized in connection with a real process, but the complete law is defined outside with all the possible properties. One of the properties of Eq. (5) is that the mass m can take an infinite value (if $v = c$). That is, a metaphysical element (the infinite) appears in connection with Eq. (5).

A theoretical world conception is formulated in terms of general equations, which allow us to describe a large variety of real systems; the law given by Eq. (5) describes a large variety of real systems varying with m_0. Such general laws are of course not presentable in space and time. Only specific systems, i.e., system with a specific value for the rest mass m_0, can be depicted in space and time. This is consistent with what we observe in everyday life.

As we have pointed out above, the infinite has to be considered as a typical metaphysical entity. It appears in many facets, not only with respect to Eq. (5). For example, an electrical field is defined quite generally by an infinite set of values. Therefore, each specific electrical field, depicted in space and time, constitutes a metaphysical element. There are a lot of other facets and faces, respectively, in connection with the infinite.

1.10.2 Where are the Non-Material Entities in the World Outside?

The organization of matter in the world outside (within space and time) is achieved on the basis of the theoretical conceptions of the world, which contain such metaphysical entities and the basic statements. Therefore, these non-material entities should somehow be positioned in material reality. These elements should also exist outside independent of the observers mind. But where have we to search, at what positions in space and time can we find them? No doubt, when we try to answer this question seriously we quickly end in a ridiculous situation.

The main problem is that the metaphysical elements cannot be positioned within the container of the real world outside. It is simply unthinkable to position the infinite somewhere in space or to assume that it occupies the entire space at each time τ. These metaphysical entities etc. cannot exist within space and time, and this is because there is no relationship between them and the space-time elements x, y, z and time τ. These metaphysical entities cannot be expressed in terms of x, y, z and time τ, and this is of course also valid for scientific laws:

$$law \neq law(x, y, z, \tau) \tag{6}$$

The same is the case for the basic statements, and it would be ridiculous to search for them in space and time. Clearly, such metaphysical elements together with the basic statements cannot be positioned inside the container; the scientific laws cannot be simply packed up within the container (Eq. (6)).

From this point of view the container principle becomes more than a questionable concept. We will recognize below that the "projection principle" seems to be more realistic when we try to order the world, in particular the "things" around us.

We may formulate the situation in this way: All these material and metaphysical entities as well as the basic statements are combined within certain mathematical laws (physical formulas). These physical laws as, for example, Schrödinger's equation or Newton's equations of motion, reflect general statements about nature and cannot be positioned in space and time. Only specific solutions of such equations for specific systems (for example, a hydrogen atom) can be described in terms of space and time [1, 2], that is, only concrete systems can occupy space and time.

1.10.3 Concluding Remarks

In the understanding of the real world we need material objects but also metaphysical entities; furthermore, for the organization of matter in space basic statements are necessary. However, for all these "things" we cannot find a classification scheme when we work within the framework of the "container principle". These things cannot be packed up within the container (space). Projection theory offers a solution.

1.11 CLASSIFICATION WITHIN THE FRAME OF THE PROJECTION PRINCIPLE

1.11.1 Peculiarities

Let us first repeat the main facts which lead to the projection principle. As we have pointed out above, all observers are caught in space and time. However, space and time, i.e., the elements x, y, z and time τ, may not appear in the reality outside because they are not accessible to observations. Thus, when we construct theoretical conceptions for the world outside, we have to find — within the projection theory — other variables; in Sec. 1.6.3 we have marked these new variables by the letters R, S, T, Q.

Because we are caught in space and time, the quantities R, S, T, Q have to be constructed on the basis of the space-time elements x, y, z and time τ. Then, we develop a world view with the help of the variables R, S, T, Q. The results have to be projected on space-time formed by x, y, z and time τ and we obtain an image (it is image A in Fig. 14). This image reflects our impressions that we have in front of us in everyday life, but also those facts which have been determined with an experimental setup based on a theoretical concept. The situation is summarized in Fig. 14 (projection principle).

As we already know, the impressions that we have in front of us in everyday life do not appear via "thought" but they come into existence spontaneously without any intellectual operation of the observer. The source for these spontaneous images, which have their seat in box A of Fig. 14, is exclusively defined by the "basic reality", which has to be considered as the "true world".

The reality which is formed by the variables R, S, T, Q cannot be considered as the basic reality but, as we have already outlined above, the variables R, S, T, Q have to be constructed on the basis of space and time with x, y, z and time τ. Thus, the reality based on the variables R, S, T, Q (it is box B in Fig. 14) is not a world independent of the observer, but it is a "fictitious reality". We have to construct fictitious realities in order to find realistic images in front of us which furthermore describe the results of measurements, also described within box A. There is no one-to-one correspondence between the fictitious reality (box B in Fig. 14) and the image (box A in Fig. 14), i.e., the image of the fictitious reality.

That is all what we can say about the world outside within the frame of the projection theory. We never can make statements about the basic, true reality that exists objectively, i.e., independent of human observers. Thus, within projection theory basic reality has to be considered as a "metaphysical system".

1.11.2 General Equations

The theoretical world conception is formulated in terms of general equations which allow us to describe a large variety of real systems. However, an image in space and time, based on x, y, z and time τ, can never reflect such a general state, because the general formula (which must be compatible with the projection principle) is principally not describable in space and time. It reflects *all* possible systems, i.e., all at once. The parameters within a general formula remain uncertain as long as we do not put into the general formula system-specific quantities. Such a "general" situation cannot be described in space and time with x, y, z and time τ.

Thus, the images in front of us, which are described in space and time, exclusively refer to concrete systems, where their properties are conformal with the general equation. That is, we put into the general formula a system-specific set of data, typical for a certain system in nature, and obtain as a solution an image represented in space and time (as, for example, Fig. 1).

As we have already mentioned, the general equations, which characterize the theoretical conception of the world, have to be compatible with the projection principle. That is, they must be formulated on the basis of fictitious realities having the variables R, S, T, Q (it is box B in Fig. 14) and the result must be projected onto space and time formed by x, y, z and time τ. It turned out (see in particular [1, 2] and Appendix F) that the variables R, S, T, Q are expressed by the momentum $\mathbf{p}$ and the energy E:

$$\begin{aligned} R &= p_x, \\ S &= p_y, \\ T &= p_z, \\ Q &= E \end{aligned} \tag{7}$$

where p_x, p_y, p_z are the usual components of the momentum $\mathbf{p}$ with $\mathbf{p} = (p_x, p_y, p_z)$. More details concerning fictitious realities are pointed out in [1] and Appendix F.

1.11.3 Real Effects

When we touch with a finger a certain object in front of us, there is merely a contact between geometrical objects; the touch takes place in space and time and there are only geometrical objects here (see Fig. 14, box *A*), that is, the human being as well as the object appear as geometrical configurations in space and time. But we "feel" this contact in space as a physically real process, and this is because everything (human with finger an object) is embedded in the basic reality as well. In other words, this physically real process takes place in the basic reality, which remains however hidden and is principally not recognizable.

The human as well as the object are positioned in the "basic reality" and an interaction process takes place in the basic reality between the material objects (here the human and the object) as soon as the finger of the human touches the object in space.

Both processes happen simultaneously. Just this process in the basic reality is felt by the observer as a certain force because it is a physically real process. However, the physically real process itself remains hidden because it takes place in the basic reality.

Since basic reality is principally not accessible, we are dependent on a "fictitious reality" when we want to describe the process theoretically. In this connection interaction means that there is an exchange of momentum and energy (see Eq. (7)) between the "material finger" and the "material object", that is, we have changes Δp_x, Δp_y, Δp_z and ΔE; the momenta and the energies fluctuate. These processes take place in the fictitious reality (as the replacement for the basic reality). In the space-time picture (Fig. 14, box *A*) no processes take place because there are merely touches between geometrical entities in box *A*.

1.11.4 Levels of Reality

As already pointed out several times, we do not know the structures within the basic reality. It exists, but we cannot make any statement about it. Even space and time do not belong to the basic reality. Nevertheless, all physically real processes exclusively take place in the basic reality without space and time. Our cognition apparatus can "only" form pictures of the basic reality.

Because space and time do not belong to the basic reality, there can be no one-to-one correspondence between the processes in the basic reality and what we experience in space and time in the form of images, which are our direct impressions in everyday life in front of us (Fig. 1). However, the theoretical conceptions of the world cannot be described within space and time; these conceptions are more general than the specific images projected onto space and time. We discussed this point in detail above, but let us mention the situation once again: Whereas a theoretical conception reflects all possible systems, the space-time image always contains specific, concrete systems as, for example, a hydrogen atom and/or other material systems. In other words, the theoretical conception is positioned on a higher level, higher than the level where the space-time images are positioned. In this way we come to the "levels of reality". All levels are constructions by the observer and belong to the brain of the observer.

These levels are arranged vertically in accordance with the degree of generality where the level with a higher generality is positioned above that with a lower degree of generality. In [1, 2] we called this kind of order the "principle of level-analysis". Here we restrict ourselves on two levels.

1.11.5 Principle of Level-Analysis

All levels of reality as, for example, level L_1 and level L_2 in Fig. 17, reflect certain features of the same world outside which we have called the basic reality. We may assign to each level certain objects, which are qualitatively different from each other. In other words, to each level in Fig. 17 belong certain "objects", i.e., geometrical objects in space (level L_1) and symbolic objects (equations), which exist as "objects" but without space (level L_2). How do we experience these "objects" at level L_2?

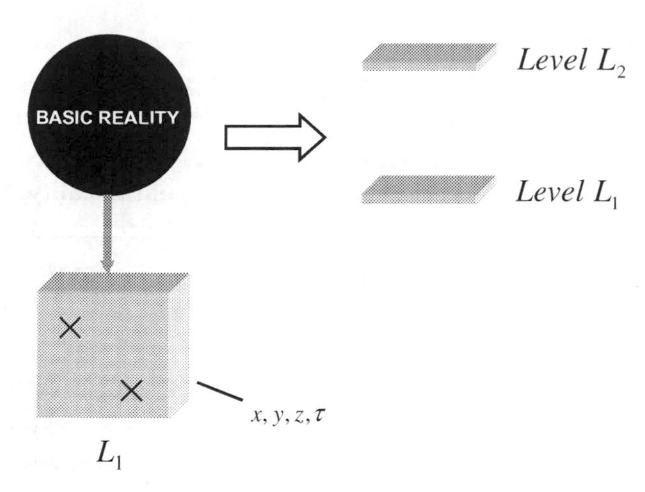

Fig. 17 We cannot make statements about the structure of the basic reality, which we have called the true reality (real world outside). However, we can observe and/or describe aspects of it within the framework of levels (levels of reality). These levels are vertically arranged in accordance with the degree of generality. This leads to the so-called "principle of level-analysis". In the figure two levels are quoted: Level L_1 and level L_2. Level L_2 is more general than level L_1. Level L_2 contains the "theoretical conceptions of the world", which are formulated in terms of general equations. Such a general formula cannot be represented by an image in space and time. A certain theoretical conception is formulated in terms of a general equation which allows us to describe a large variety of real systems. However, an image in space and time, based on x, y, z and time τ, can never reflect such a general state, because a general formula reflects all possible systems (all at once) and such a situation can principally not be described in space and time. Only specific systems (hydrogen atom, etc.), where their properties fulfil the general equation, are depictable in space and time (level L_1, box L_1). Clearly, the images that we have directly in front of us (Fig. 1 is an example) within assumption-less observations in everyday life are depictable in space and time too and belong also to level L_1 and box L_1. Box L_1 is positioned on level L_1.

Level L_1 can be considered as the "material level" because we "feel" the geometrical objects, positioned in space, as forces (see Sec. 1.11.3). These geometrical (material) objects come into play through the spontaneous observations in everyday life and also by observations with the help of measuring instruments.

The description of all these space-time configurations at the material level (level L_1) have to be done by means of the "fictitious realities" introduced above that replace the basic reality using the variables p_x, p_y, p_z and E (see Eq. (7)); more details are given in Fig. 14. The entire information in fictitious reality is then projected onto space and time [1, 2].

On level L_2 we have the symbolic objects (equations). What is here the effect on the human observer? How do we experience such a symbolic object? Like the geometrical (material) objects on level L_1, the symbolic objects on level L_2 are in the brain of the observer and reflect certain properties of basic reality. But the human observer cannot simply interact with a "formula" as it were a material object. In Sec. 1.11.3 we discussed such an interaction with a material object; here we investigated a human finger which touches a geometrical object. However, the situation with objects on level L_2 is quite different from that. We cannot touch a formula with our fingers because such formulas are not representable in space and are therefore not material in character. That is, we cannot apply the conception of fictitious realities characterized by the variables p_x, p_y, p_z and E (see Eq. (7)) and, therefore, we cannot have fluctuations Δp_x, Δp_y, Δp_z and ΔE that we feel as physically real forces on our body. An object on level L_2 as, for example, a physical law in the form of a formula, cannot be grasped and experienced in this way. Here non-material interactions are necessary.

But how do we "feel" an object positioned of level L_2? In other words, how does a human "feel" a formula? Once again, in the case of the finger and an object of level L_1, the finger interacts materialistically with the object, and the human body feels the properties (form, inertia, etc.) of this level L_1-object as forces expressed by the fluctuations Δp_x, Δp_y, Δp_z and ΔE, which formally take place in fictitious reality. Let us mention once again, that the physically real process actually takes place in the basic reality; fictitious realities are a certain kind of auxiliary realities for the description of the geometrical structures in space and time.

Such a "materialistic interaction" is not imaginable when we try to grasp the "object" formula, which is a level L_2-object. Instead we can find a connection to the formula by an intellectual operation (whatever "intellectual operation" may mean here), and the intellect recognizes the features of the formula by certain operations.

To sum up, we interact with the formula (physical law) intellectually. This is a strictly non-material process and takes place in the basic reality. We cannot use fictitious realities here because this kind of reality describes the geometrical structures in space and time and are therefore exclusively responsible for material processes. This in particular means that intellectual realities are not representable in space and time. Other frames are needed here that are abstract in character.

The "intellectual operations" at level L_2 correspond — in a certain sense — to that what we experience at level L_1 in the form of "forces". Whereas the intellectual operations are conscious actions, the forces come into existence unconsciously. We "feel" the physical law through conscious operations with the law itself. We intellectually interact with the physical law.

1.11.6 The Objects of Projection Theory

The "principle of level-analysis" allows a straight classification with respect to the notions real bodies (matter), real processes, metaphysical entities and what we have called the "theoretical conception" of the world. Real matter is exclusively positioned in the basic reality; also the real processes and real effects exclusively take place in the basic reality. However, the basic reality is principally not observable and, therefore, it has to be classified as a "metaphysical system". We cannot know the structure inside the basic reality; here the real structure as well as the theoretical structure are synonymous. On the other hand, the direct impressions in front of us — in the observations of everyday life and also specific systems (hydrogen atom, etc.), which are specific solutions of general equations — are exclusively geometrical structures that are positioned in space and time (level L_1); these geometrical structures have their seat in the brain of the observer. The "general equations" are also positioned in the brain of the observer and belong to level L_2 (Sec. 1.11.4 and Sec. 1.11.5). These equations are constructions of the observer. Again, the information on level L_1 and on level L_2 are "pictures and symbols" in the brain and reflect certain peculiarities of the basic reality.

There is a clear line between the metaphysical system (basic reality) and what appears as a picture in the brain (see Fig. 18). The entities

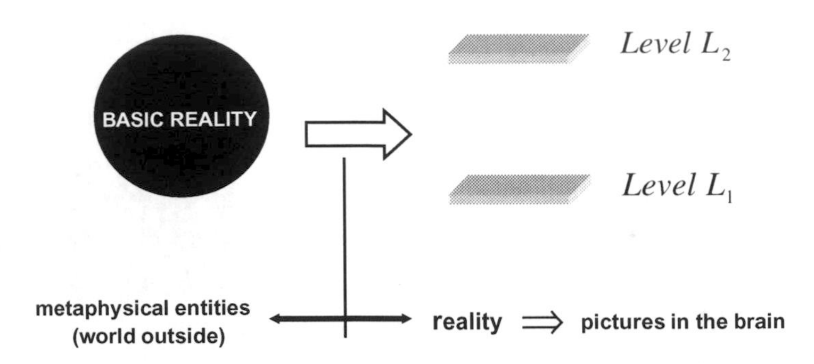

Fig. 18 Real bodies (matter), real processes, metaphysical entities and that is what we have called the "theoretical conception" of the world can be classified within the "principle of level-analysis". Real matter is exclusively positioned in the basic reality, furthermore the real processes and real effects exclusively take place in the basic reality. However, basic reality is principally not directly observable because a picture-independent point of view is not defined for a human observer. Therefore, the basic reality has to be classified as a "metaphysical system". We can principally not know the real structure and also not the theoretical structure inside the basic reality. But what do we know? We experience the world on certain "levels of reality" (level L_1 and level L_2 in the figure).

Level L_1: Here the direct impressions in front of us in the observations of everyday life and specific systems (hydrogen atom, etc.), which are specific solutions of general equations, are exclusively geometrical structures which are positioned in space and time, i.e., on level L_1. These geometrical structures have their seat in the brain of the observer.

Level L_2: The "general equations" are also positioned in the brain of the observer and belong to level L_2. (Sec. 1.11.4 and Sec. 1.11.5) and also reflect certain peculiarities of the basic reality. These equations are constructions of the observer. Again, the information on level L_1 and on level L_2 are "picture and symbols" in the brain. There is a clear line between the metaphysical system (basic reality) and what appears as a picture in the brain. The entities which appear on both levels have to be considered as real objects, although they are not material in character, we experience them in connection with physically real effects and intellectual operations. The appearance of formulas and other physical statements is not mysterious; all these things belong to the brain, no less and no more. Everybody knows that.

which appear on both levels have to be considered as real objects, although they are not material in character; we experience them in connection with physically real effects and intellectual operations. Within the projection principle the appearance of formulas and other physical statements is not mysterious. All these things belong to the brain, no less and no more.

Here no "mixed states" consisting of material objects and metaphysical entities appear as in the case of the container principle (Sec. 1.8). In fact, the problems seem only to appear when we apply the container principle that is connected in conventional physics with the pretension that one is able to find the "absolute truth"; it is very often believed that the so-called "world equation" can be found soon. This is however a fallacy when we work on the basis of the projection principle. From the point of view of the projection principle the "absolute truth" (basic reality) is principally not accessible to a human observer, but this principle also says that it definitely exists.

1.11.7 Levels of Observations

From the various levels of reality (level L_1 and level L_2, Fig. 17) emerge properties that are qualitatively different from each other although the various levels reflect features of the same reality, i.e., the basic reality. We may state that from each level emerges a certain facet of the basic reality, a facet of the world outside.

The difference between the "objects" on the various levels is not only reflected in their theoretical description but also — as we have recognized — by their effect they have on the human observer. Thus, we may assign to each level of reality a "level of observation" (see Fig. 19) because the features of the "objects" on the various levels are defined differently, that is, the methods of observations must vary from level to level and these produce certain level-specific feelings inside the observer. We feel the objects on level L_1 (trees, cars, etc.) differently from those on level L_2 (theoretical conceptions).

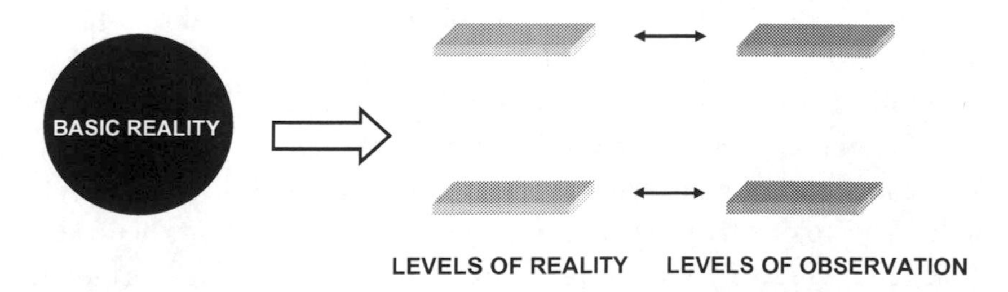

Fig. 19 Levels of reality in relation to the levels of observations. The features of the "objects" on the various levels are defined differently and, therefore, the methods of observation must be different from each other. The correspondence between the levels is essential.

The objects at both levels are positioned in the brain of the observer. However, there is a big difference: On level L_1 we observe unconsciously and on level L_2 consciously by thinking.

1.11.8 On the Structure of the Basic Reality

The basic reality should be considered as a "unified whole" and not as a large system consisting of separate things, which are qualitatively different from each other. All aspects experienced and/or defined by the human observer do not exist in a separated form in the basic reality. The various levels of reality are constructions by the human being and belong to the brain. All levels reflect certain features of the same world outside (basic reality). Thus, the basic reality should be considered as a "unified whole" without levels separated from each other. Separation is in particular also a peculiarity at the material level where the objects appear as geometrical objects in space and time. Here separation is a feature due to the existence in space and time, but in the basic reality there is no space and time and no such separation.

Mind, matter and what we often call the "soul" belong to specific aspects positioned on various levels of reality, but should not exist in this separate form in the basic reality. Instead the features such as mind, matter, soul, etc., should exist in the basic reality as one (unified) state. In the analysis of the structure of the basic reality a holistic view and not the separation into parts or levels would be appropriate if we were able to recognize details of the basic reality, but we are not.

In other words, the differentiation between mind, matter, etc., is not a suitable concept when we think about the nature of basic reality. All these things exist in basic reality as a unified block and one should not be able to distinguish between matter, mind, etc., even when in the basic reality individual features are established. However, this unified block appears in the "observer's world" as a system of various levels. This in particular means that this feature is dictated by the observer's peculiarities.

Why is the world organized in this way? Why do we have not the absolute truth directly in front of us? These are difficult and basic questions. The correct answers are given by biological evolution, and we will recognize this peculiarity when we discuss this point in Chap. 2.

1.12 THE PROJECTION PRINCIPLE VERSUS THE CONTAINER PRINCIPLE

Within the projection theory the world is recognizable only with the help of images, which are positioned on various levels. There is a level where the material objects are positioned (level L_1 in Fig. 17) and there is a level where the intellectual structures (formulas etc.) are placed (level L_2 in Fig. 17). However, all these "objects", which belong to the various levels, are states of the brain; metaphysical entities do not appear because all these objects cannot directly belong to the basic reality, and only the basic reality can be considered in the projection theory as a metaphysical system.

Metaphysical entities are metaphysical in character because they are defined but not accessible to observations, although they give rise — together with real matter — to physically real effects if they belong to the world outside (container principle). Just this is the critical point. Equations for the description of the space-time images with all their real and metaphysical objects should not belong to the world outside. For details, see Sec. 1.10.

This is not fulfilled within the container theory because within this concept the structures in the images are identical with those in the world outside where the container is made of space and time. This point has been discussed above, in particular in Sec. 1.10.

The situation is different within the projection theory. Here we have a "basic realty" and "levels of reality" with certain objects, but all these objects on the various levels should only be exist in connection with space and time (material level) and/or as abstract pictures (intellectual level with the theoretical conceptions). The objects on all these levels are exclusively positioned in the brain of the observer.

Since basic reality should be considered as a "unified whole" and not as a large system consisting of separate objects, the specific objects positioned at the levels of reality should not appear in the basic reality, particularly since space and time cannot exist in the basic reality. From this point of view it is not possible to transfer the objects from the levels to the basic reality and vice versa. The elements (entities) in the basic reality are not defined in the way the brain does. Such a correspondence is even unthinkable. This underlines and justifies our above made statement about the basic reality: It is not observable and has to be considered as a large metaphysical system. However, the objects on the various levels are not metaphysical in character, because none of them can give rise to physically real processes and effects. For example, the elements of space and time x, y, z and time τ do not belong to the basic reality and can therefore not give rise to physically real processes. Thus, the elements x, y, z and time τ should not be considered as metaphysical in character. Such statements only make sense when we work within the projection theory.

Again, all aspects experienced and/or defined by the human observer should not exist in a separated form in the basic reality; instead, the basic reality should be considered as a "unified whole". The information, which goes from the basic reality to one of the levels, is processed, i.e., it is selected and transformed.

Within the frame of the container principle we construct a reliable theoretical conception of the world outside and define its structure through the elements of this theoretical conception. Then, all the elements in the formulas of the theoretical construction are positioned within the world outside, i.e., within space and time. As we already outlined several times, the direct observation of the world outside is here principally not possible. Within the frame of the container theory the "world outside" is de facto identical with the "basic reality", i.e., the container itself with all the embedded things is considered as the basic reality in conventional physics.

Once again, in the theoretical conception of the world outside not only physically real entities appear but also non-observable metaphysical elements as well as other basic statements, which are also not accessible to observations. On the other hand, these non-observable metaphysical entities influence the configuration of the material world outside significantly, i.e., the metaphysical entities operate like physically real objects but are not observable. Such a concept is however not acceptable from the scientific point of view. This is the situation when we work within the framework of the container principle.

One of the difficulties within the container theory is that the metaphysical elements cannot be positioned within the container of the real world outside, that is, within space and time. This point has been expressed by Eq. (6) in Sec. 1.10.2.

The main problem is that the space-time elements x, y, z, and τ, forming the container, are metaphysical in character: Although they are not accessible to observations they are considered to be responsible for physically real effects; for example the effect of inertia within Newton's mechanics and also in the Theory of Relativity. Such non-scientific peculiarities cannot be accepted.

The projection principle (Fig. 14) is based on another concept; here the basic reality works without space and time, i.e., without the elements x, y, z, and τ. Let us underline once again that the quantities x, y, z, and τ are not accessible to empirical tests but behave, on the other hand, as physically real entities within conventional physics. This is a non-acceptable situation.

All these metaphysical entities together with space and time, characterized by x, y, z, and τ, which are metaphysical as well, behave like physically real quantities in the world outside when we work within the container principle. All these quantities have to be dissociated from the members of the world outside, i.e., they have to be moved from the world outside to the image (box A in Fig. 14, which is identical with box L_1 in Fig. 17). This is fulfilled within the frame of the projection theory.

Such metaphysical elements and basic statements are not only in the "mind of the observer" but belong particularly to the "world outside" itself; they should exist without the observer within the frame of the container principle. The organization of matter in the world outside in space and time is also here achieved on the basis of such metaphysical elements and basic statements and, therefore, they should somehow be positioned in a reality independent of the observer's mind.

In summary, within the frame of the container principle we have "mixed states" in the world outside consisting of real matter, metaphysical entities and other basic statements, which essentially determine the configurations in the world outside. The container itself with all the embedded things is considered in conventional physics as the basic reality.

1.13 THE SHAPE OF THE OBSERVER IN THE BASIC REALITY

What can we say about the shape (outer form) of a human observer in the basic reality? We are all familiar with the observer's shape in space and time, but we can nothing say about it in the basic reality. The reason is obvious: Space and time with its elements x, y, z, and τ do not exist in the basic reality and, therefore, a picture of a human observer in the basic reality remains principally hidden (see also Fig. 20). This must have serious consequences: Since the basic reality reflects the absolute truth, the observer is not able to recognize the absolute truth about himself, i.e., about his true shape and of course the true nature of all the other things.

Since the structures in the basic reality are principally not accessible, we cannot conclude from the space-time structures the form of those structures which are existent in the basic reality, which is symbolically expressed in Fig. 21.

1.14 THE PRODUCTS OF IMAGINATION

The principle of level-analysis refers so far on the laws and operations in the physical realm. Here the physical laws are effective, and the world is considered from the point of view of natural sciences. The situation is discussed in Sec. 1.11.4 and Sec. 1.11.5 — in this connection Fig. 17 is relevant. On level L_1 the "hard" objects (material bodies) are positioned which appear in front of us as geometrical

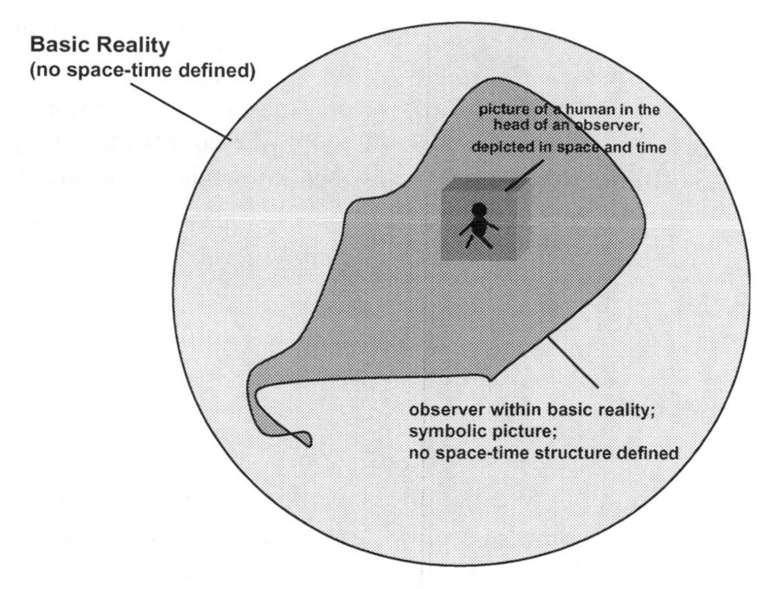

Fig. 20 A human observer is able to make statements about himself (his shape) only in connection with space and time, but he has not the ability to conclude from this space-time picture, that what is existent of him in the basic reality. This is of course valid for all the other things. Therefore, the form of the observer in the basic reality have been chosen arbitrarily in the figure; this arbitrarily chosen shape expresses that the space-time picture of the observer in his head must be different from that which is existent of him within the basic reality.

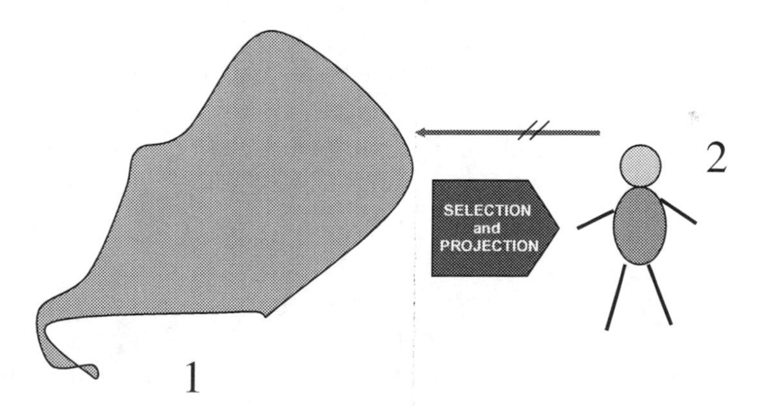

Fig. 21 The observer in the basic reality is here characterized by structure 1. His shape has been chosen arbitrarily because we can say nothing about the observer in the basic reality; a picture-independent point of view is not possible within the projection theory (more details are given in the text). The observer, positioned in space and time, is characterized by the familiar structure 2 as it appears after a selection and projection process from the basic reality. The observer is principally not able to conclude from structure 2, familiar to all observers, the form of structure 1 (observer in the basic reality).

objects. These hard objects are organized in space and time on the basis of the physical laws, which belong to level L_2.

The material bodies, i.e., their geometrical pictures in space and time, are exclusively impressions of our brain and appear spontaneously in front of us (Fig. 1 is an example). However, there are other "things" in our brain that do not obey physical laws. Let us call them the "products of imagination". What are the peculiarities of such appearances?

1.14.1 Thoughts

In order to be able to answer this question, we would like to consider first a simple example. We may for example imagine that a car moves from the earth to the moon in a split-second. Such and similar thoughts exist but do not obey the physical laws: The material body "car" cannot move from the earth to the moon because the construction does not allow that. Furthermore, the car is not able to move such a large distance in a split-second because this situation would not be conformal to the elementary physical laws, in this case the Special Theory of Relativity.

On the other hand, these "strange" thoughts are in the mind like the material bodies around us. These thoughts are, in other words, as real as the material bodies are real entities in our world.

However, such "products of imagination" are qualitatively quite different from that what we experience in our assumption-less observations in everyday life, which are positioned at level L_1 in Fig. 17.

The spontaneous images in front of us (Fig. 1) are geometrical structures in space and time. Also those (material) things that we observe with our measuring instruments, which are described on the basis of fictitious realities, are assigned to the same space-time, i.e., to the same space-time we experience automatically within the framework of our assumption-less observations in everyday life. Essentially, all material bodies and material processes have exclusively their seat in this space-time.

However, the "products of phantasy" ("products of mind") do not appear in this space-time; we do not observe them spontaneously or with measuring instruments together with the "material bodies". Thus, the products of phantasy exclusively appear in connection with another space and time, it is the space-time of our imagination and is not that of our observations that are positioned on level L_1 (Fig. 17).

Since the products of phantasy (imagination) are qualitatively different from the material entities and processes, they cannot be placed on level L_1; they have nothing to do with level L_1 and also not with level L_2, because they do not obey in general the physical, scientific laws and do not represent physical laws.

The products of phantasy belong to the brain. Here the general form of the brain is meant, i.e., the brain of the basic reality. We are only familiar with the material part of the brain, that is, its space-time structure, which obey the physical laws. A lot of scientists actually believe that this material part of the brain is able to create that what we call the "mind" as, for example, the products of phantasy. This has to be considered as a fallacy when we work within the frame of the projection theory.

How can a system create certain products, which are not compatible with the system's internal laws? This should not be possible; such products have to be in accord with the systems internal laws.

We cannot derive the products of phantasy from the physical laws. This is almost trivial because these products do not belong to the class of appearances that obey the physical laws (theoretical conceptions of the world, positioned on level L_2). So, the material part of the brain, obeying the physical laws, cannot create the products of mind (phantasy) because these products do not in general obey the physical laws. A car may move in our phantasy (in our thoughts) from the earth to the moon in a split-second. Such and similar thoughts exist but do not obey the physical laws.

In a nutshell, the products of mind (phantasy) can obviously exist without the material part of the brain; the source of mind is obviously not of the material level.

However, we know that there must be a certain connection between "mind" and "matter", between the "products of mind" and the "material states" within the brain. If there were no such a correlation between mind and matter, we would come to the Cartesian division. Within the Cartesian division mind and matter are considered as two separate parts, which are independent from each other. However, this division created a serious problem: If man's will, which can be considered as a result of an operation of the mind, has no contact with the matter of its body, how can it compel this body to turn to the right or to the left as it pleases?

To summarize, there is a certain connection between "mind" and "matter but — and this is important to say — matter does not produce the products of minds. Due to the correlations between mind and matter, the products of phantasy (and all similar things) may be influenced by matter but cannot be produced through matter. In other words, if the material part of the brain is changed (for example by an accident and/or in connection with a medical operation) the mind can be changed too but it cannot be created or annihilated in this way.

1.14.2 Siegfried Lenz and Other Writers

These statements are in accordance with the opinion of many writers. In a book "About Imagination" (Über Phantasie" [14]) the writers Siegfried Lenz, Heinrich Böll, Günter Grass, Walter Kempowski and Pavel Kohout discuss the role of imagination in literature and life. We do not want to discuss the details here, but we would underline that these writers believe to know that imagination (phantasy) is possible without experience, i.e., imagination can exist, in their opinion, without the world of the five senses. This is completely in accordance with the result of the projection theory: The products of mind can exist independently of matter; the products of mind are not created by the material processes in the brain (its material part).

That the products of phantasy are frequently of limited use can be considered as natural. In analogy to the material things and the physical laws, the products of phantasy are produced in large numbers, from which however only those are selected, which are useful and which are compatible to those things that are already existent. As it is well known, this is exactly the situation with the material bodies at the biological level. They are formed according to the principles of evolution. Nature puts a vast amount of genetic material to the test, in order to select the small portion that fits into existing conceptions; the remainder is rejected, as in the case of phantasy. Whereas the products of phantasy are selected consciously, the biological material is selected unconsciously. Essentially, there is a certain analogy between the "products of phantasy" and the "products of matter". Nevertheless, in most cases only the products of matter are considered to be useful.

Even if the imagination has become discredited because it is often regarded as excessive, it must be said that on the other hand human life is not conceivable without phantasy and imagination, respectively. In this connection the following statement from Meyer's encyclopedia is of interest: *Imagination as a specific human ability is involved in all perceptions, actions and plans. In particular, productive thinking and creativity are not conceivable without imagination.* No doubt, this is an instructive statement.

1.15 SUMMARY OF THE MAIN STATEMENTS

1. No doubt, the observations in everyday life are of basic relevance. We normally base our analyses on those events and things which are spontaneous in front of us (see, for example, Fig. 1). We experience these events and things directly without any interpretation. In such situations we have humans, trees, cars, the sun, the moon, stars and a lot of other things directly in front of us and they appear without any intellectual help.

We grasp the material world within the framework of assumptionless observations. This world appears to be embedded in space, and reality is considered as a "container" in which all the material bodies are positioned. We have called this concept the "container principle". All traditional physical theories work on the basis of this principle.

However, the impression in front of us reflects an inner state within the brain of the observer. The problem is that we have no direct access to the world outside.

How is this "inner world" related to the "world outside"? The simplest assumption is that there is a one-to-one correspondence between the structure in the image (Fig. 2) and the structure of the world outside (Fig. 3), and we come to the container principle.

Within the container principle, the world outside is embedded in space and time. But space and time also belong to the observers mind. This in particular means that space and time are tightly linked to the biological system (human being).

In other words, the space (space-time) appears twice within the container principle, in the inside world as well as in the world outside, i.e., there are two space-time types. This is however problematic and more than questionable. Why? One of the reasons is the biological evolution leading to the following feature: The appearance of space and time within the observer's brain means that the features of the inner space and the inner time are influenced or even constructed by the processes of biological evolution.

Clearly, the features of space and time of the "world outside" cannot be influenced by such factors, i.e., by the biological evolution. Reality outside is assumed to be independent of the observer's peculiarities. In other words, within the container principle two space-time types appear with different peculiarities. This however reflects an inconsistency.

2. The structure of the world within assumption-less observations reflects a certain level. The levels on which we develop theories and interpretations are in general more sophisticated and also more detailed. However, also at these scientific levels we work within the container principle. All theories about the world, from Newton's mechanics, the Theory of Relativity to all forms of conventional

quantum theory, have been based on the container principle. We may state that the container principle is considered in conventional physics as a *basic, absolute truth*. We analyzed this situation and came to the conclusion that this standpoint is obviously a fallacy.

We in particular recognized that the material world outside should not be embedded in a container, i.e., in space and time but should be projected onto space and time and we came to the "projection principle". Here in this principle there is no space-time in the world outside.

3. What are the peculiarities of space and time? The basic characterization of space as, for example, our three-dimensional space, is given by three real numbers x, y and z (coordinates). On the other hand, time is characterized as well by points, and we have marked them by the Greek letter τ. If space and time are physically real quantities, the following question is essential: Are these basic quantities, i.e., x, y, z and τ, accessible to empirical tests? This is definitely not the case. The following facts show that there is no possibility for that.

We definitely cannot see, hear, smell, or taste single elements x, y, z and τ of space and time, i.e., the basic elements of space and time, characterized by x, y, z and τ, are not accessible to our senses. This is independent of the character of space (space-time), whether it is absolute or non-absolute. Also measuring instruments for the experimental determination of the space-time points x, y, z, and τ are not known and even unthinkable.

One might object that we experience space in everyday life (Fig. 1). However, this kind of experience has nothing to do with the observation of the basic elements x, y, z and τ. We can only say something about "distances in connection with masses", and time intervals in connection with physical processes.

Then, we come to the following important conclusion: Since we are principally not able to "observe" the basic elements of space and time (i.e., x, y, z and τ), they should never be the source for physically real effects as, for example, inertia.

This is not fulfilled within Newton's theory, but also not within Einstein's theory (Appendix B). The velocity of the single mass embedded in space (represented in Fig. 5b) remains constant ($\mathbf{v}_1 = const$) and this property is due to a physically real effect: There must be a relationship between Newton's absolute space and the mass m_1, i.e., a certain interaction between both entities (space, mass) has to be postulated, producing, in the opinion of Newton, the effect of inertia. This is however a problematic and ill construction because the elements of space-time are not observable and can therefore not be considered as a physically real something. Such a construction has to be declined.

Similar ill situations come into play within the General Theory of Relativity; this is quoted and discussed in Appendix B. Within the Special Theory of Relativity we have a situation similar to that in Newton's theory. Here Newton's absolute space is merely replaced by an absolute space-time.

4. Again, we never observe single elements x, y, z and τ, but we are only able to observe "distances in connection with material bodies (masses)" and "time intervals in connection with physically real processes". This in particular means that the elements of space and time, characterized by x, y, z and τ, are not defined as physically real entities and are therefore not observable. Such a space-time block can never be the source of physically real effects because it does not have any physically real existence. What are the consequences?

"Matter" and "space" are closely linked; neither should be able to exist without the other. This point is important and we analyzed this point in more detail. Let us consider two bodies, say 1 and 2, having the masses m_1 and m_2 and the coordinates x_1, y_1, z_1 and x_2, y_2, z_2 at time τ. Our observations are always done within space. Matter is embedded in space, and we never observe that matter leaves space or enter it. This in particular means the following: The space coordinates x_1, y_1, z_1 and x_2, y_2, z_2 can only exist in connection with the masses m_1 and m_2 (Eq. (1)); the masses m_1 and m_2 can only exist in connection with space coordinates, i.e., in our example we need x_1, y_1, z_1

and x_2, y_2, z_2. Since only distances in connection with "two" material bodies (masses) are observable, a single, isolated elementary body cannot be observed and is therefore not a real physical entity. In other words, a single elementary body cannot exist in space.

Let us go a step further and put into the same space "two" elementary bodies (body *A* and body *B*), and we would like to assume that there is no interaction between the two bodies with m_A and m_B. Then, body *A* is not existent for body *B*, and body *B* is not existent for body *A*. In other words, body *A* behaves like a unit as if it would be alone in space; the same is true for body *B*. Therefore, both bodies cannot exist as in the case of a lone body in space.

Essentially, body *A* and body *B* behave like isolated entities that are non-existing. What we have said in connection with "one" single body in space is also valid for the configuration consisting of body *A* and body *B*: Both bodies *A* and *B* are non-existing entities. Because body *A* and body *B* are not defined, we are not able to define a distance between body *A* and body B. An empty space remains. However, an empty space, characterized by an infinite ensemble of coordinates x_i, y_i, z_i, without bodies does not exist.

We assumed that there is no interaction between the two bodies *A* and *B*. However, such a configuration may not exist. Therefore, we not only need the space coordinates $x_A, y_A, z_A, x_B, y_B, z_B$ and the masses m_A and m_B of the two bodies, but there must in addition a "relation" between them, and such a relation is expressed by an interaction between body *A* and body *B*. This interaction leads to correlations between the coordinates, so that distances become definable.

If both elementary bodies interact, they are able to exist in space and a distance between them can be defined. Only such kind of configuration may exist because only such kind of system is able to fulfill what is observable in nature and what is scientific. We can only say something about "distances in connection with masses" (and time intervals in connection with physical processes).

What about the "interaction"? It is, as we have pointed out, necessary for the construction of a realistic physical reality. What kind of interaction is required? It is an interaction, which produces the elementary bodies itself, that is, body *A* produces body *B* and body *B* produces body *A*. It is an existence-inducing interaction, and must be independent on the distance between the bodies.

In a nutshell, "existence-inducing interactions" are necessary. This is a quite general statement and is independent of the conception, i.e., whether we work within the "container principle" or within the "projection principle".

Within conventional physics (Newton's mechanics, quantum theory, the Theory of Relativity, etc.) we work within the "container principle" and distance-independent interactions are not known and would be difficult to introduce. That is, "existence-inducing interactions" can hardly be introduced within the frame of the "container principle". However, within "projection theory" [1, 2] the world is not embedded in space and time and all the points (in connection with inertia, existence-producing properties etc.) can be treated without any problem. Isolated, non-interacting elementary systems are forbidden to exist in the projection theory; this is fulfilled in the most natural way. Just our basic statement about that what is observable (we can only observe "distances in connection with masses") requires that isolated elementary bodies may not exist.

5. The only physical property on which we can base our knowledge about space and time is given by the following fact: We can only say something about "distances in connection with masses", and "time intervals in connection with physical processes"; only these features are observable. Since the basic elements of space and time (i.e., x, y, z and τ) cannot be observed (defined), space and time should never be the source for physically real effects such as, for example, inertia.

We may give the following general statement: Because isolated space-positions are not existent, a body cannot be defined relative to space, but only relative to another body.

Nevertheless, the phenomena of space and time are existent. Everybody knows intuitively what space (time) is. Our world in front of us appears as space-container filled with real objects. The same is true for time: We all believe to know what time is but, however, when we try to understand the "nature of time" we normally evade this question. We feel and observe real objects, but the basic elements of the container, the space-time points x, y, z, and τ, in which these objects are embedded, may not to be considered as physically real. No doubt. This seems to bc a contradiction. In this monograph we propose a solution for this problem.

Again, the world in front of us appears as space-container filled with real objects (Fig. 1). When we touch with our fingers certain objects (tree, car, etc.) we definitely feel them, i.e., the objects and the observer's body interact with each other. Both, the objects as well as the observer, are considered as physically real objects. We make this statement on the basis of the facts of the inside world, but there are no material objects within this "inside world" (symbolically expressed by Fig. 2). This is however no problem because it is normally assumed that there is an exact "one-to-onc-correspondence" between the reality outside and the inner picture of it. Exactly this concept is used in conventional physics.

6. When we observe the world in everyday life, an image of it appears directly in front of us. However, we do not have objects in front of us that are embedded in space and time having the elements x, y, z, and τ. We merely observe "objects" and "extensions". For example, two objects (geometrical positions in the image) have a certain "extension". Here the notion "extension" has to be considered as a basic notion and we should not try to analyze it further. "Extension" reflects a qualitative effect. The effect of "extension" appears spontaneously in front us in connection with our assumption-less observations in everyday life, i.e., it appears without thinking. We have a lot of bodies in front of us and, therefore, we have a lot of extensions. The brain organizes this ensemble of extensions as *one* phenomenon which we call "space". That is all what we can say about space. All that leads to a qualitative impression of space. But we need more than qualitative impressions when we want to analyze the physical processes in the world theoretically.

It turned that space and time and its basic elements x, y, z and τ can only appear in the inner world and they do not belong to the elements of the world outside. Then, the coordinates x, y, z at time τ are exclusively the elements of a "fictitious net" which the observer intellectually puts over the image in front of him. Thus, space and time, constructed in this way, can never be the source of physically real effects. This in particular means that the basic elements x, y, z and τ are exclusively entities of the brain because they come into existence by thinking.

Since space and time do not exist in the world outside, there can be no one-to-one correspondence between the facts in front of us and the world outside. A one-to-one correspondence is only given within the container principle.

7. How do we get an idea of the structure of the physical world outside? There is basically only one possibility: We create a physical conception of the world outside by thinking and this gives rise to questions, i.e., questions that are put to nature itself. That is to say we carry out specific experiments and the deflection of the pointer on the measuring instrument is the answer to our questions.

This is the only way to construct configurations for the world outside. Actual reality outside is in principle not directly accessible to an observer, and this is because we are not able to take an image-independent point of view, that is, there is no external point of view which would enable a direct observation of the world outside. We never recognize a theoretical conception spontaneously without thinking.

Three things are relevant when we try to assess the observer's relation to the world outside.

1. The world in front of us, the image of the reality outside, is the most important fact that the observer can have about the reality outside. This image is a configuration in space and time.
2. The second point is that space and time cannot be entities of the real world outside, and we can say nothing "directly" about this reality,

that is, about the true (basic) reality; this world is not accessible to an observer because an image-independent point of view is not possible.

3. We are only able to say something about the "reality outside" "indirectly" with the help of theoretical conceptions, i.e., on the basis of intellectual imaginations. These theoretical conceptions have to be checked with experimental instruments.

8. All observers are caught in space and time. Since space and time, i.e., the elements x, y, z and time τ, cannot not appear in the reality outside, we have to construct other variables when we try to construct theoretical conceptions for the reality outside. In this way we obtain a "fictitious reality". We have to construct fictitious realities in order to find realistic images in front of us and which describe the results of measurements. There is no one-to-one correspondence between the structures in fictitious reality and those in the image in front of us. That is all what we can say about the world outside within the frame of the projection theory. We never can make statements about the basic, true reality that exists objectively, i.e., independent of human observers.

9. Again, all observers are caught in space and time. However, space and time, i.e., the elements x, y, z and time τ, may not appear in the reality outside because they are not accessible to observations. Thus, we have to find — within projection theory — other variables; we have marked these new variables by the letters R, S, T, Q.

The quantities R, S, T, Q have to be constructed on the basis of the space-time elements x, y, z and time τ. Then, we develop a world view with the help of the variables R, S, T, Q; the results have to be projected on space-time formed by x, y, z and time τ and we obtain an image, which reflects our impressions of what we have in front of us in everyday life, but also those facts which have been determined with an experimental setup based on a theoretical concept.

The impressions we have in front of us in everyday life come into existence spontaneously without any intellectual operation of the observer. The source for these spontaneous images is exclusively defined by a "basic reality", which has to be considered as the "true world", which is however not accessible to human beings. Within the projection theory the basic reality has to be considered as a metaphysical system.

The fictitious realities, described by the variables R, S, T, Q, cannot be considered as the basic reality. Once again, there is no one-to-one correspondence between fictitious reality and the image, that is, the image of fictitious reality.

10. The "theoretical conceptions" of the world cannot be described within space and time, and this is because these conceptions are more general than the specific images projected onto space and time.

Whereas a theoretical conception reflects all possible systems, the space-time image always contains specific, concrete systems as, for example, a hydrogen atom or other material systems. In other words, the theoretical conception is positioned on a higher level, higher than the level where the space-time images are positioned. In this way we come to "levels of reality". All levels are constructions by the observer and belong to the brain activities of the observer.

All levels of reality reflect certain features of the same world outside (basic reality). We may assign to each level certain objects, which are qualitatively different from each other. In other words, to each level in Fig. 17 belong certain "objects", i.e., geometrical objects in space (level L_1) and symbolic objects (equations), which exist as "objects" but without space-time (level L_2).

11. What is the shape (outer form) of a human observer in the basic reality? We are all familiar with the observer's shape in space and time, but his outer structure in the basic reality remains hidden. The reason is obvious: Space and time with its elements x, y, z, and τ do not exist in basic reality and, therefore, a picture of a human being in the basic reality is not accessible (see also Fig. 20). What we can say with certainty is that there can be no one-to-one correspondence

between the structures in space and time and those in the basic reality. This must have serious consequences: Since the basic reality reflects the absolute truth, the observer is not able to recognize the absolute truth about himself, i.e., about his true shape and of course the true nature of all the other things.

Chapter Two

THE IMPACT OF EVOLUTION

In the preceding sections an essential question was in the centre of our considerations: Is the world in front and around us the basic, objective reality? In other words, is what we observe in everyday life within the frame of assumption-less perceptions the absolute truth? The indications that it is not are overwhelming. In the basic reality, cars, flowers, sun, moon and the stars cannot have the structure as they directly appear to us, that is, as they appear spontaneously before our eyes.

In the last section we quoted essential reasons why the impressions before us cannot be the absolute truth. However, we did not discuss the deeper cause for this peculiarity. The facts that the basic reality remains hidden to the human observer are obvious, but we do not know yet why nature is organized in this way. Let us try to find the deeper reason for nature's concept.

Once again, how can a human being not be expected to recognize the absolute truth about the things of his environment? The answer is given by the strategy of nature, and the plan of this strategy in connection with biological systems is primary to "survive" and not to recognize the basic, objective reality. This feature is clearly reflected in the principles of evolution, i.e., humans and other biological systems developed over the course of time in accordance with the condition for survival and not to recognize the absolute truth. Biological evolution is essentially based on "usefulness", and the "principle of usefulness" is valid here. Reality outside is not assessed by "true" and "untrue" but by "favourable towards life" and "hostile towards life". This strategy essentially determined the development of the recognition apparatuses of man and of course those of other biological systems.

2.1 THE INFLUENCE OF EVOLUTION

All these are weighty and relevant statements, and this view requires further evidence and discussion. The inclusion of evolution is a fundamental aspect and should have in general a repercussion on the structure of the basic physical laws. If man's recognition apparatuses developed according to the features "favourable towards life" and "hostile towards life" and not by "true" and "untrue", then the structure of the physical laws, constructed by human beings on the basis of "his" recognition apparatuses, must be influenced by the same evolutionary conditions.

In particular, our theoretical conception of the world outside should not lead to "absolute" statements. In fact, we already pointed out in Chap. 1 that there can be no one-to-one correspondence between the structures in the image, which we have directly in front of us, and the structures in the basic reality. That is what we can say with certainty, although we do not know the structure of the basic reality.

A relevant question arises: How do we have to assess to the structure of the present laws of physics under these conditions, dictated by evolution? We already discussed this topic in [1] and [2] on the basis of a more mathematical analysis. In this book we will continue and deepen that, but without specific mathematical considerations.

The problems addressed in this monograph are by no means only academic in character, but it is absolutely necessary to know the relationship between the observer's perception and reality outside. For example, in nanoscience and nanotechnology it is planned to change or to improve the brain functions of biological systems. Such manipulations will in general be dangerous, and we have to model the specific processes that take place in the brain "before" we start such experiments.

The process, which produces the "world view" of the individual comes into existence within the brain (for details, see Chap. 1), and this process must be consistent with the corresponding model construction [15]. If that should not be the case, the individual could face problems in mastering his environment and even survival is in general not guaranteed.

No doubt, a theoretical model for the brain will be influenced by the basic structure of the laws of physics. Such a brain model has to be reliable when we want to change certain brain functions by nanotechnological means. It is however difficult to assess the reliability of the present laws of physics if their structure is dictated by evolutionary conditions. What do we expect?

In Chap. 1 we already gave some general statements concerning the relationship between the world outside (basic reality) and the image in front of us, which is produced by the brain. It turned out that there can be no one-to-one correspondence between the structures in these images and the structures in the basic reality. We will recognize that the cause for all these facts and peculiarities is the strategy of nature (evolution).

2.2 INFORMATION

Let us mention once again that man's perception of that what we call "reality" is essentially determined by the principles of evolution, in particular by the principle of usefulness. According to this characteristic, human beings developed adhering to the following criterion: He only selects the information from the environment that is "useful" for him, which is useful for life.

2.2.1 An Important Point: As Little Outside World as Possible

In Chap. 1 we came to the conclusion that a human being is not able to observe the basic reality, i.e., he is principally not able to say something about the "absolute truth". It is quite clear that the impressions in front of us, experienced in everyday life, do not reflect the true reality. Nevertheless, this experienced world could contain the complete information of the universe, at least in principle. Since the experienced world is essentially relevant for the development of theoretical conceptions, also these theoretical developments could be considered as complete if we were able to observe a complete world. This is however not the case. Why?

The perception of complete reality in the sense of a precise reproduction implies that with growing fine structures increasing information of the outside world is needed. Then, the evolution would have furnished the sense organs with the property to transmit as much information from the outside world as possible. But the opposite is correct: The strategy of nature is to take up as little information from the reality as possible. Reality outside is not assessed by "complete" and "incomplete" but by "favourable towards life" and "hostile towards life". Concerning this point, Hoimar von Ditfurth stated the following [16]:

No doubt, the rule 'As little outside world as possible', only as much as is absolutely necessary is apparent in evolution. It is valid for all descendants of the primeval cell and therefore for ourselves. Without doubt, the horizon of the properties of the tangible environment has been extended more and more in the course of time. But in principle only those qualities of the outside world are accessible to our perception apparatus which, in the meantime, we need as living organisms in our stage of development. Also our brain has evolved not as an organ to understand the world but an organ to survive.

The perception of reality by biological systems is essentially influenced by the principle "as little outside world as possible". On the other hand, this principle can be understood by means of the idea of evolution.

The principle of evolution, i.e., the phylogenetic development from simple, primeval forms to highly developed organisms, can be considered as the key for the development of biological systems, in particular their way of how they perceive the world outside. It is the theory of evolution by natural selection which is generally accepted in the meantime. Its foundations have been created by Charles Darwin (1809–1882) more than one hundred years ago. Since then it has been modified and developed further by geneticists. Evolution by natural selection is a two-step process: *Step 1*: By recombination, mutation, etc., genetic variants are produced at random. Populations with thousands or millions of independent individuals arise. *Step 2*: Some of these independent individuals will have genes which enable them to manage the predominating situation due to the environment (climate, competition, enemies) better than others. Thus, they have a larger chance for survival than others. They will have, in the statistical average, more descendants than other members of the population. Natural selection takes place in favour of those organisms where their genes have adapted to better cope with the environment.

The number of examples that biological systems have developed in accordance with these criteria is overwhelming. Man and other creatures are characterized by this species-preserving appropriateness. The principle "as little outside world as possible" is compatible with the principles of evolution; it is a succession of evolution. Only those things that are useful for a human and other creatures are relevant. Therefore, the principle "as little outside world as possible" is in a certain way a "principle of usefulness". In essence, the goal of evolution was not to develop creatures (biological systems) to recognize the "absolute truth", but to treat reality outside with respect to "favourable towards life". Thus, from the point of view of evolution it is quite natural that a human being is not able to have access to the basic reality. Instead we have to work within so-called "fictitious realities". We have discussed this point in detail in Chap. 1, in particular in connection with Fig. 14.

2.2.2 Information Transfer

The principle "as little outside world as possible" refers to the evolution of the material part of reality. How all the other "things" as, for example, the "products of phantasy", are influenced by evolutionary effects, remains at first an open question.

The basic reality contains the complete information about the world. Only a part of it is projected on space and time, and this follows directly from the principle "as little outside world as possible". In Chap. 1 we have called the part that is projected onto space and time the "material reality". This material part is defined through our spontaneous impressions, which we have directly in front of us in everyday life, and of course by the facts we observe with suitable measuring instruments.

In connection with observations, we are caught in space and time but also with respect to our thoughts. Therefore, the theoretical conceptions of the world, which we construct intellectually, have to be considered as dependent on the state of evolution, which a certain biological system (for example a human being) occupies. This is almost a matter of course because we want to describe that information from the basic reality, which has been selected adhering to the principle "as little outside world as possible". Thus, all our theoretical views in physics only represent restricted information about the material world, which is projected onto space and time.

If the total information in the basic reality is C and if A is a part of it, which is projected onto space and time, then we have in any case $A < C$, and we are principally not able to say something about that part of C which is not needed by the observer. In other word, the physical laws that transform and select the information from the basic reality onto the space-time frame in front of us, are principally not accessible (see in particular Chap. 1). This also means that we cannot map A on C. This situation is schematically represented in Fig. 22.

Because of this futile situation we introduced in Chap. 1 the notion of the "fictitious reality". A fictitious realty is a certain kind of world outside for the theoretical description of the material reality, defined as a projection on space and time. Just this part is reflected in a fictitious reality. Let us denote the information within fictitious reality by B. The theory shows that the information within fictitious realities is exactly the same as those given in the space-time images, i.e., we have $A = B$.

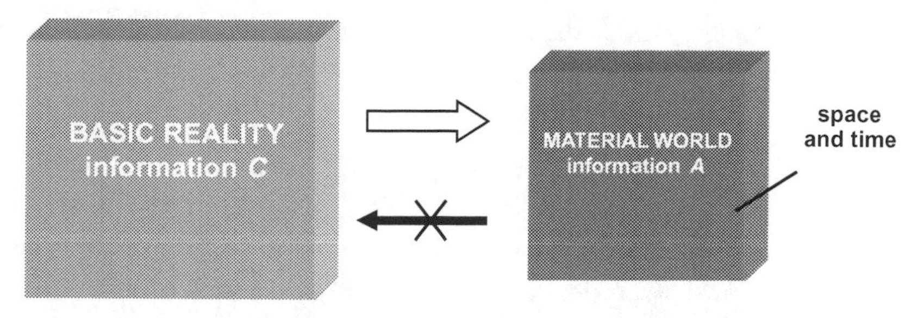

Fig. 22 According to the principle "as little outside world as possible", only a part A of the complete information C about the world is transformed from the basic reality to the space-time image, which reflects material reality. Since we have always $A < C$, we are principally not able to say something about that part of C which is not needed by the observer. We cannot conclude from information A anything about information C. The physical laws, which transform and select the information from the basic reality to our material world in front of us, are principally not accessible.

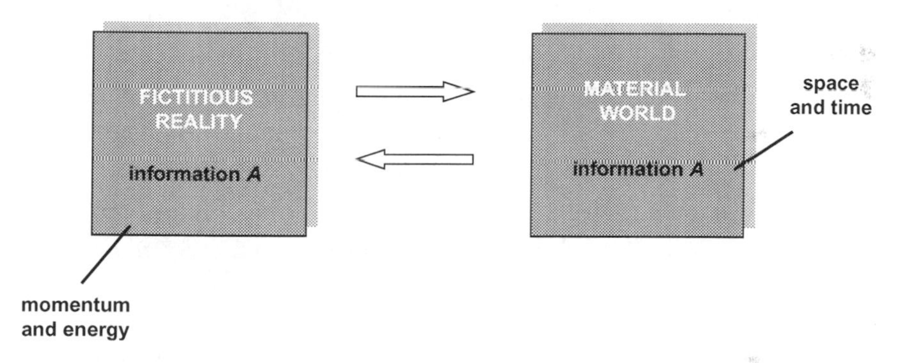

Fig. 23 Because the basic reality remains hidden, we construct fictitious realities. The information within fictitious realities is exactly the same as those given in the space-time images (which reflect the material world), i.e., it is A. The laws, which transform the information of fictitious reality on space and time and vice versa, are known and are given by the Fourier transformation.

The laws which transform the information of fictitious reality on space and time and vice versa are known and are given by the Fourier transformation (see [1, 2] and Appendix F). The situation is summarized in Fig. 23.

The variables of fictitious reality are expressed by the momentum $\mathbf{p} = (p_x, p_y, p_z)$ and the energy E (see Secs. 1.6.3, 1.11.2, 1.11.3 and Appendix F). The interaction takes place within fictitious reality in the form of $\mathbf{p}$, E — fluctuations Δp_x, Δp_y, Δp_z and ΔE. Exactly this information is projected (transformed) onto space and time. Clearly there can be no interaction processes within space and time.

2.2.3 Material Realities

What we called "material reality" is a definition. How is it defined? This is a relevant question and needs to be answered explicitly. The reason is obvious: Material reality represents an important factor in physics and is the basis for certain philosophical directions such as materialism.

The material reality is defined through our "spontaneous" impressions, which we have directly in front of us in everyday life and of course by what we observe with suitable measuring instruments. The material world experience in this way consists of the bodies around us as well as of atoms, molecules, elementary particles and other objects that we cannot directly observe by means of our eyes.

Material realities appear as "geometrical structures", which are exclusively "representations in space and time", whereby space and time are characterized by the elements x, y, z and time τ.

This is however a construction (definition) by the human observer. In other words, we should not assume that the material entities appear as separate units in the basic reality as well. The bodies around us as well as what we call atoms, molecules, elementary particles etc. do not reflect an "absolute" fact. As we have already outlined in Sec. 1.11.8, the basic reality should be considered as a "unified whole" and not as a large system consisting of separate things, which are qualitatively different from each other. All aspects experienced and defined by the human observer should not exist in a "separated form" in the basic reality.

Since material reality is experienced by the human observer "unconsciously", we have the impression that these material entities exist independently from the observer, that is, we consider them as objectively real. Nevertheless, what we call material world is only a definition through the "human observer".

On the other hand, thoughts and in particular the products of mind etc. come into existence by "conscious" activities and, therefore, these human peculiarities (thoughts, etc.) are felt to be dependent on the observer, and most people consider them as "subjective". Thus, mind, which creates all these things, is very often believed to be considered to be produced by the material reality: The subjective fact is believed to be created by the objective fact. We have pointed out several times that this belief is obviously a fallacy.

We marked the total information in basic reality by C and due to the principle "as little outside world as possible" only the selected part A is projected on space and time, which we defined as "material reality". We always have $A < C$, and we can principally nothing say about the part $\alpha = C - A$. Also the laws, which transform and select the information A from the basic reality to our world, remain principally unknown. In this way the "material reality" of the human observer is defined.

Let us analyze the situation in more detail, and let us mark the human observer by S. We have to assume that for another kind of observer the principle "as little outside world as possible" is valid as well. We will mark this type of observer by S'. Also S' selects "spontaneously" from the basic reality (having a constant information content C) a certain part of it, say A'.

As in the case of S, the material reality of S' is defined as follows (quite in analogy to that what we defined above for S):

The material reality of S' is defined through the "spontaneous" impressions, which S' has directly in front of him in his everyday life and of course by what he observes with suitable measuring instruments. The material world experienced in this way consists of the bodies around S' as well as certain entities (analogous to the atoms, molecules, elementary particles) and other objects that S' cannot directly observe by means of his sense organs.

Since we have assumed that S' is different from S, the information A' should be different from A, i.e., we have $A' \neq A$. The information

A' defines the "material reality" of the other type of observer. Also here "material reality" means that it appears as a "geometrical structure" within the brain of S', which is projected onto a frame, which is different from space and time since S' is different from S. That is, it is not the space-time elements x, y, z and time τ that define the projection frame of S' but other elements, say $a, b, c, \ldots$.

As in the case of S, we always have $A' < C$, and S' can principally nothing say about the part $\alpha' = B - A'$. Also the laws, which transform and select part of the information A' from the basic reality to the world of S', remain principally unknown.

Essentially, the material reality of the other type of observer is different from the material reality of the human observer:

$$(material\ reality)_{S'} \neq (material\ reality)_S \tag{8}$$

That what we call "material reality" is observer-dependent. It is a construction (definition) by the observers S and S'. Again, it is probable that the material entities do not appear as separate units in the basic reality. There might be no information overlap between both the material realities. Then, S does not perceive S' and vice versa. This situation is illustrated in Fig. 24. The bodies around a human observer S as well as what he calls atoms, molecules, elementary particles etc. do not belong to the material reality of S', the other type of observer. In other words, S' defines his own material reality on the basis of other entities. If there is a certain information overlap we come to Fig. 25. S perceives a certain part of S' and vice versa.

To sum up, the term "material reality" does not reflect an "absolute" fact. It does not appear in the basic reality but is obviously a strict observer-dependent definition. The material entities defined by S are different from those defined by S'.

2.2.4 Constructed Realities

The principle "as little outside world as possible" determined the physical development of the individual and how one learns to handle other individuals and also the environment. In particular, the "unconscious world view" of the individual, that is, the picture of the world

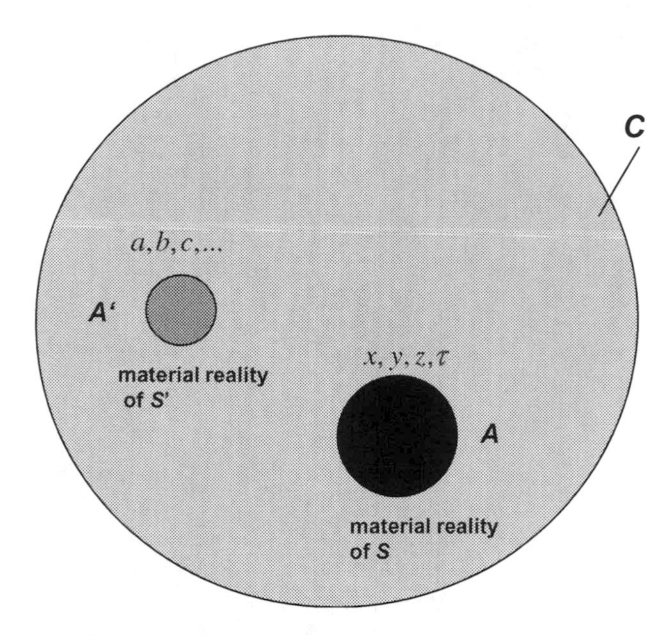

Fig. 24 There is no information overlap between the human observer S and the other type of observer S'. A' defines the material realty of S', which is projected onto the frame with the elements a, b, c, ... A defines the material realty of S, which is projected onto the frame with the elements x, y, z and time τ. The large sphere C is the total information in the basic reality.

which appears unconsciously in front of him, is fixed by this kind of specific information. However, due to the evolutionary selection process this representation must be an incomplete view of the reality outside. In other words, there can be no one-to-one reproduction between what we have in front of us and what is actually positioned in the reality outside. An example for this incomplete impression of the world outside is represented in Fig. 1.

All the "products of mind" and the "products of phantasy" etc. do not appear in such images, but have to be considered as real as the images in front of us; both types of appearances are likewise states of the brain and reflect in particular certain facts of the basic reality. The products of mind also reflect certain features of the basic reality, but they are positioned on another level than the material objects and cannot be depicted within space and time. The situation is analogous to what we have discussed in Sec. 1.11.5, where we introduced level L_2

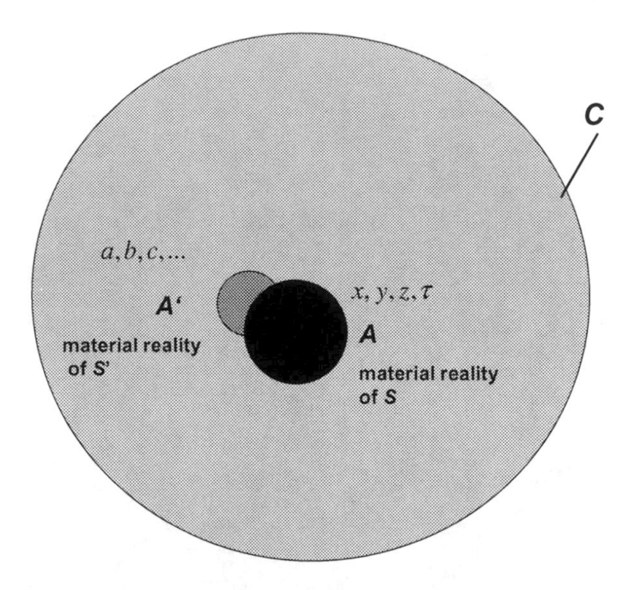

Fig. 25 There is a certain information overlap between the human observer S and the other type of observer S'. A' defines the material realty of S', which is projected onto the frame with the elements $a, b, c, \ldots$. A defines the material realty of S, which is projected onto the frame with the elements x, y, z and time τ. The large sphere C is the total information in the basic reality.

on which the "theoretical conceptions of the world" are positioned (see in particular Fig. 17). The situation in connection with the "products of mind" should be quite similar.

No doubt, the images in front of us, i.e., the material world, are in the centre of all our actions. But the mind is essential too. The intellectual, mental and spiritual features, developed by individuals, consciously create information that can be used by the individual for the improvement of life conditions.

Both the unconscious and the conscious states have exactly the same task: They determine the possibilities to cope with the outside reality, in particular to construct favourable life conditions. Thus, the unconscious and the conscious creation of information are equivalent in character; they are different from each other but do not differ principally. Both states have the task to determine the relationship between the individual and his environment.

We may state that the "unconscious world view" is extended by the "products of mind" leading to an extended world view of man. In this way we obtain a "constructed reality".

No doubt, this is an unusual conception because most people are firmly convinced that the impressions before them (assumption-less observations in everyday life) belong directly to a basic, objective reality, which reflects what we normally call the "material world". The "products of mind" are normally considered as less relevant. However, due to the principles of evolution this conception is obviously questionable. We have no access to the actual (material) reality outside, but the unconscious and the conscious states are equally positioned in the brain of a human being, and this is probably valid for all biological individuals.

Even when the "unconscious world view" is (almost) the same for all individuals, the "constructed world view" (extended world view) is in general different for different human beings, that is, it varies from individual to individual because each individual has his own world of ideas and thoughts, respectively. This point will be discussed further and deepened in the sections below.

2.3 NO PRINCIPAL DIFFERENCE BETWEEN MATTER AND MIND

So far, the essential point of our considerations is the notion of "information" and its role in connection with our view of reality. How the information comes into existence, consciously or unconsciously, is not the relevant factor in the assessment of the various elements of information.

Again, the objects at the material level, which we observe unconsciously in everyday life, have in principle to be considered as equivalent to the "objects" produced consciously at the level of mind. In other words, within this conception of the world there is no principle difference between matter and mind. Clearly, we experience matter differently from the "products of mind", but this point is not the relevant factor.

Due to the strategy of nature, expressed by the laws of evolution, we can principally not know how the basic, objective reality is formed and what its information content is. We have only access to a small part of reality outside, and we cannot evade this situation. To construct the complete reality outside from the restricted part of it is not possible because we have to assume that the entire information embedded in reality outside is not compressed within the small, selected part we observe; selection and evolution, respectively, would not make sense.

2.3.1 Paul Watzlawick

This view of the world is confirmed by specific investigations in the field of communication research; here the notion "information" is of basic relevance.

The ideas developed so far agree largely with those developed by the psychologist and philosopher Paul Watzlawick (1921–2007). Watzlawick was also a communication researcher and was in particular interested in the origin of information and its use in life. He believed that the reasons for interpersonal conflicts are rooted in the fact that we have an incorrect idea of reality. His interesting ideas can be summarized as follows [17]:

People stumble again and again into conflicts because they assume that they know what the basic, objective reality is. Therefore, everybody believes that they recognize the ideas of others. However, this is obviously not the case since everybody lives in his own reality. It is a reality constructed by men. It is a "constructed reality" as in the case we introduced in Sec. 2.2.4 in connection with unconscious and the conscious information. Within Watzlawick's conception "constructed reality" means that man ascribes a specific meaning to certain situations. Or in the words of Watzlawick [17]:

I am in my own reality, just as you are in your reality. We naively assume that there is an objective reality. This however is not correct. If you ask me which reality am I in, then I will tell you I am in the reality constructed by myself, that is, I give the situation now and here a specific meaning. If you give the situation a basically different meaning we have an interpersonal conflict. Then the problems start.

On the basis of this idea, measures and solutions can be found which help to avoid or to eliminate conflicts. Watzlawick discusses these solution procedures. In [17] we find among other things an interesting comment:

... A good idea, suggested once by the logician Anantol Rapport, would be for example, if the following procedure were used between the world powers during all talks: Before the negotiations are started, each delegation must explain satisfactorily the point of view of the other delegation to that delegation.

Thus the Americans ought to try to explain the Soviet point of view to the Soviets in such away that the Soviet say: "Yes, that is correct, that is the way we see things." And then the Soviets have to explain the point of view of the USA so that the Americans say: "Yes, that is our point of view." If this were done, presumably half the problems would already be solved before they were even discussed.

This discussion shows that the basic, objective reality is not recognizable and that we have to seek recourse to constructed realities, even outside the realm of physics. This is particularly reflected in the fact that a single situation can be perceived as having different meanings indicating that there is a certain insecurity, and an absolute standpoint is not recognizable in such and similar cases.

According to Watzlawick an objective reality does not exist. From our point of view objective reality exists but remains hidden. Or rather, it can really be said that for an observer objective reality does not exist, at any case not immediately.

The main problem is to find a realistic conception for the relationship between the observers perception and what is actually outside. This is obviously not only a problem of physics but, as Watzlawick demonstrated, also a point in psychology. It is astonishing that psychology and physics (in connection with the principles of evolution) come exactly to the same result (see also Fig. 26): The basic, objective reality cannot be recognized. We have in any case "constructed realities".

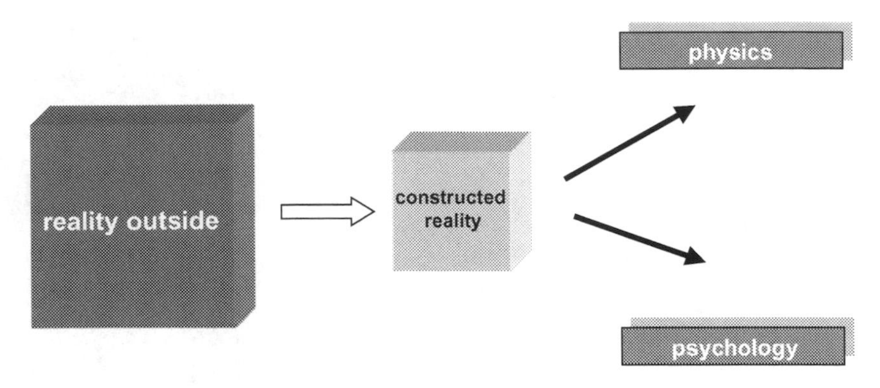

Fig. 26 The relationship between the reality outside and perceived reality is of fundamental relevance in physics as well as in psychology. The fact that we cannot recognize basic, objective reality is equally important for physics and psychology (conflict research).

This relationship between "reality outside" and "constructed reality" has been deduced within the frame of physical arguments but, as Watzlawick demonstrated, is obviously also reflected in conflict research (psychology).

Conclusion

Due to the successful application of this conception of constructed realities in conflict research as well as in physics, we should consider it as a general principle, valid for all possible situations in the world. It has to be applied for example when we want to change certain brain functions nanotechnologically. We may state that the conception of "constructed realities" will be reflected in many other situations.

2.3.2 Connections

As we have already outlined in Sec. 2.2, the "unconscious world view" is (almost) the same for all individuals. However, the extension of this unconscious perception of the world by the "products of mind" leads to the "constructed world views", which are in general different for different human beings, i.e., it varies from individual to individual because each individual has his own ideas and thoughts, respectively.

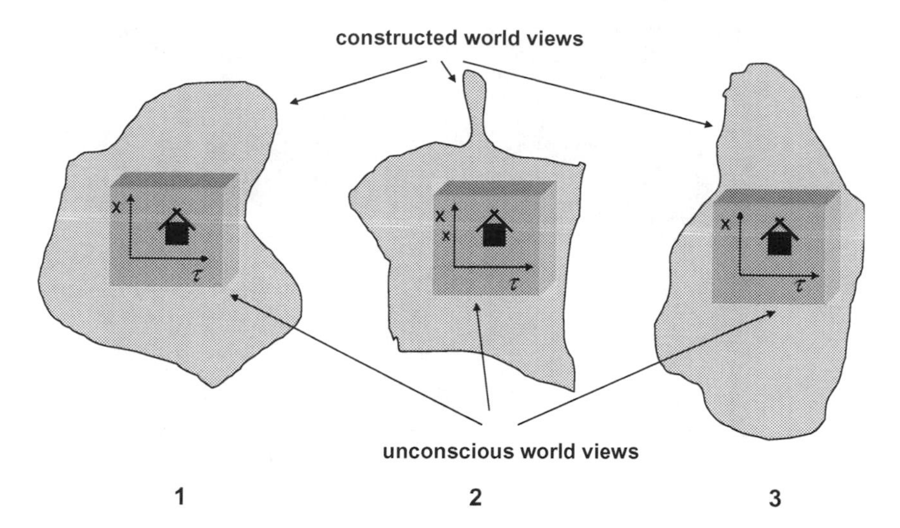

Fig. 27 The constructed world view is different for different human beings; world view 1 is different from 2 and 3. However, we know from experience that the unconscious world view is (almost) the same for all individuals; it is given by our direct impressions in front of us, which we experience spontaneously in everyday life. This direct world view is extended by the "products of mind" and we obtain constructed world views.

This is summarized in Fig. 27. In Fig. 27 we recognize how the constructed world view is obtained: The "unconscious world view" is extended by the "products of mind" leading to an extended world view for man. In this way we obtain a "constructed reality". In Fig. 28 the constructed world view is identical with construction 2 given in Fig. 27. Concerning conscious and unconscious constructions we have essentially two situations:

1. *Conscious constructions*

The observer develops consciously specific ideas and thoughts on the basis of his mind. These specific conscious constructions could be used for example to change the material world, in particular for the improvement of the life conditions.

2. *Unconscious construction*

The selected information from reality outside is used for the unconscious construction of a picture that appears spontaneously in front of us.

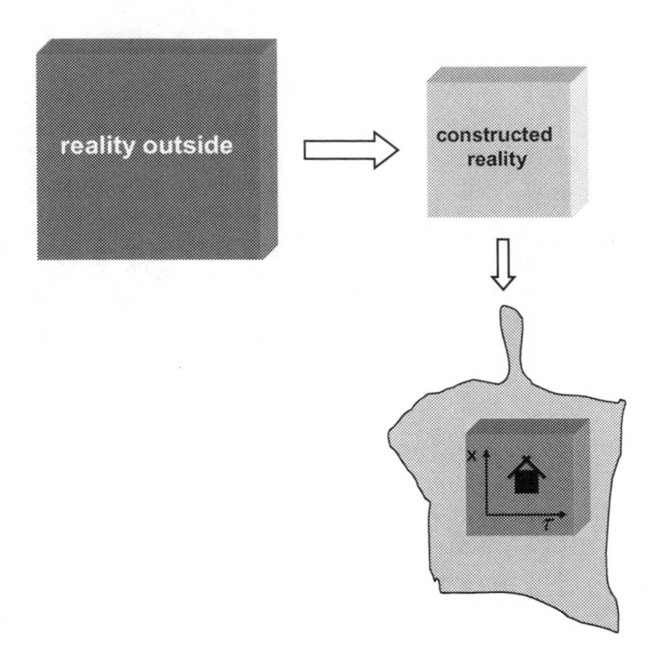

Fig. 28 The constructed world view is identical to the construction given in Fig. 27 (in this case for individual 2).

Classification of Watzlwick's view

Watzlawick (Sec. 2.3.1) argued very convincingly that the roots for interpersonal conflicts are due to the belief that our ideas and thoughts are absolute in character. It is obviously often believed that the specific views of a human being should also be valid for other human beings.

What is the reason for this questionable view? We may suppose that it is due to the assumption that the certainty, which is undoubtedly given at the unconscious level, is also valid for the conscious part of the constructed reality. In other words, the constructed realities are assumed to be (almost) identical for all individuals and, therefore, the three constructed realities in Fig. 29 are given by exactly the same representations. We know that this view is a fallacy.

2.4 TRADITIONAL VIEWS

What peculiarities do we connect with the notion "mind"? Let us answer this question on the basis of the abilities that we normally

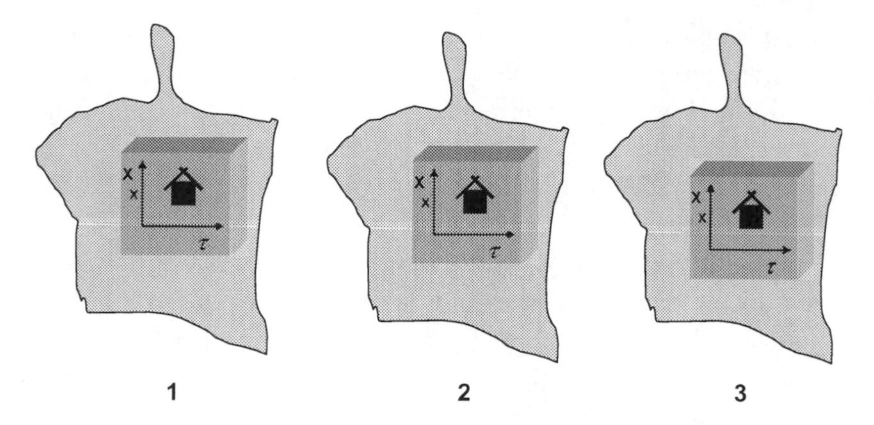

Fig. 29 In the opinion of Watzlawick (Sec. 2.3.1) the roots of interpersonal conflicts are due to the erroneous conception that the ideas and thoughts of a human being are absolute in character, that is, a human being obviously believes that his views should also be valid for other human beings. What is the reason for this questionable view? We may suppose that it is due to the assumption that the certainty, which is undoubtedly given at the unconscious level, is also valid for the conscious part of the constructed reality. In other words, the constructed realities are assumed to be (almost) identical for all individuals (reality 1 = reality 2 = reality 3) and, therefore, the three constructed realities in this figure are given by exactly the same representations. This is however a fallacy as we know.

expect a mind to be capable of: The mind of a human being enables thinking, consciousness, reasoning, perception, judgement etc. However, these features may also apply to other biological systems, may also be valid to other life forms.

The mind-body problem is a central and most interesting point in connection with the nature of mind. Is what we call "mind" an independent unity, i.e., separate from specific physical phenomena as, for example, from neurological processes? Or is the mind the result of material processes within the brain? This question has been answered quite clearly in Chap. 1 within the frame of the projection principle: The mind is not a creation of the material part of the body of human beings. On the other hand, traditional thinking has more or less led to a materialistic view.

2.4.1 Features of Mind

Quite independent from the question whether there is a relationship between the mind and the physical body, it is generally accepted that the mind enables subjective awareness and intentionality with respect to the environment and to other individuals.

No doubt, there are uncertainties with respect to the definition of the notion of the "mind". This is particularly reflected by the fact that many different cultural and religious traditions treat the concept of the mind in very different ways.

We would like to demonstrate here that traditional thinking has led to the view that the mind seems somehow derived from physical phenomena. This standpoint is mainly based on the belief that the derived physical laws (as for example Newton's theory) reflect the "absolute truth". In other words, with the development of the physical laws the mind and its peculiarities is often considered to be derived from physical phenomena, that is, the "products of mind" seem to be reduced to them. But we have to be careful.

A lot of people contradict this thesis without being able to give definite and convincing arguments against this materialistic view. However, there is obviously a way out of this dilemma when we include the phenomenon of evolution. In this section we have already applied "evolution" on the notion of "reality" and we came to the conclusion that we deal in life mostly with "constructed realities". Also the phenomenon of evolution has led us to the conclusion that the structure of the reality outside is not recognizable, and this fact has already been found in Chap. 1 without having to consider evolution.

However, history has written another book. Let us first examine this historical, traditional point of view which supposes at all levels that the structure of reality outside is recognizable.

2.4.2 The Influence of Newton's Theory and Kant's Requirement

It is usually assumed that there is a one-to-one correspondence between the structures of the world outside and what is in the picture in front of us. Most people and even scientists work on the basis of this "one-to-one principle". Why has this conception never been considered with more care? The famous philosopher Immanuel Kant made clear that such a "one-to-one principle" should be ruled out, but the scientific community ignored this requirement by Kant (more details are given in Appendix D).

The reason for that is the large success of Newton's physical theory. Newton's theory is based on our everyday life observations and considers the world around us as the absolute truth, as the basic (objective) reality. The success of his theoretical developments was so large that the starting point of his thoughts — the basic, objective reality — was really not scrutinized.

However, already at the beginning problems arose in connection with the interaction between the material bodies and even Newton himself wanted to give up his theory. But the large success swept critical questions under the carpet. Newton's theory is successful up to the present day and is still applied extensively.

Let us continue our discussion on the pre-scientific era. How did people understand the world in those days? In particular, how did people define the "products of mind" and what was their definition in connection with the material world?

2.4.3 Mythological Beings

The meaning of the objects in front of us

The world before our eyes is from elementary relevance. These immediate optical impressions, which appear spontaneously before us without conscious actions of the observer, are of fundamental importance for the interaction of a human being with the world. To all objects — trees, houses, plants, animals and generally all that is recognizable — a human being has important relations and they particularly define his existence in the physical, but also in the spiritual respect. No theory, no matter how sophisticated it is, can have continued its existence if it cannot explain the immediate world before us, that is, the objects of everyday life and their interplay with each other.

Then, for a person, who strives for recognition, important questions come up: How is this world before us to be judged? Which kind of reality reflects it? Is it the absolute or basic reality, the original world itself, or does the impressions before us only reflect a subordinated, second-class form of reality?

What kinds of things occupy space?

Before man began to argue scientifically he usually explained the phenomena in nature by means of mythological beings. It was supposed — practically without exception — that the space serves not only as a container for the material objects, but also for certain non-material things and objects, respectively, which we call today "mythological beings". These mythological beings were gods, demigods, fairies etc.

The point is that such metaphysical objects (fairies, etc.) had in those days a certain definite meaning and function, and have been considered as really existing in space, like the material objects (trees, houses, etc.) that we observe in everyday life. So, people ascribed a certain outer form to these metaphysical objects similar to those we experience optically in everyday life.

In other words, these metaphysical objects (mythological beings) were seen in the pre-scientific era as objects with actual existence that occupy space. However, they were "merely" "products of mind", but this situation implies in particular that the "products of mind" had a large rating before the scientific method was born.

In a nutshell, in the pre-scientific view not only the material objects occupied space, but the metaphysical objects (as gods, demigods, fairies and a lot more) as well.

In the opinion of the people who lived during the pre-scientific era, these metaphysical objects had an essential influence on the events in the world. So, for example, the movement of the earth around the sun was actively managed by metaphysical beings as, for example, by self-willed fairies.

In other words, moody fairies and other metaphysical objects worked in the field and in the woods and between the sky and the earth. This was the understanding, a human being had of "reality" before Galileo Galilei and Isaac Newton entered the scene: It was a space (container) which was filled with material and metaphysical objects. This world was understood as absolute in character and represents the pre-image for what we have marked as the "basic reality".

2.4.4 Intuition

Let us underline once more that — before the scientific method was born — it was usual to explain phenomena in the world by means of mythological beings: A vast number of gods, demigods, capricious fairies etc. governed arbitrarily, and they were responsible for all phenomena in the field and the forest. Statements were clothed in the form of myths, using metaphors and symbols, poetic images, and similes. The exploits of these mythological beings were the subject of fantastic tales, collected in epics of large dimensions [18].

The scientific method pushed these mythological pictures of the world away; natural science has basically changed man's view in connection with reality. In particular, modern physics has had a strong influence on almost all aspects of human society. Modern physics and its application to technological questions has fundamentally changed the conditions of life. Not mythological pictures but rational knowledge became the basis for our world views and values.

Rational knowledge belongs to the realm of intellect, which discriminates, divides, compares, measures and categorizes. Abstraction is an essential feature of the scientific method, because the comparison and classification of the things around us does not allow one to take into account all their features, but we have to select a few significant ones. In this way we develop intellectual views of the world.

The method of abstraction is very efficient and powerful, but we have to pay a price for it. As we proceed in the abstraction process, the system of concepts becomes increasingly dettached from the real world and, therefore, certain aspects of reality can be lost. For example, we do not burst out in liberating laughter at a funny joke by intellectual analysis of that joke, but by its holistic understanding. In a split-second one experiences a certain enlightenment; such a moment must come spontaneously without any intellectual action.

No doubt, there are phenomena in our world which cannot be described explicitly, but exclusively by intuition. A sudden intuitive insight helps us to grasp certain phenomena and not their understanding via a mathematical formulation. However, such phenomena are of no use to natural science and in particular to physics. Intuitive wisdom is not of specific interest in the development of technological systems and it has no particular rating in our culture. In western culture rational knowledge is favoured over intuitive wisdom. Our feelings and thoughts, our values and attitudes are influenced by this tendency; it is particularly reflected in our political and social structures.

This tendency has certainly its roots in the mechanistic world view, which has dominated our culture in the last centuries. This view is without any doubt responsible for the fact that most people favour rational knowledge over intuitive wisdom. The reason for that is the tremendous success of the mechanistic world view.

2.4.5 Some Basic Features of the Mechanistic World View

Natural science is based on the concept that the observed effects are driven by lawfulness and not by capricious fairies and/or other mythological beings. According to this concept it is assumed that the phenomena around us can be expressed by scientific laws, that is, in terms of theoretical conceptions, expressed by mathematical equations. Within Newton's mechanics the planets are objects of inert mass, moving by universal laws; they were no longer to be feared or even consulted as powerful objects which influence human affairs.

The mechanistic world view is based on Newton's mechanics, i.e., on Newton's view of the cosmos. In this regard two points are essential:

1. The objects of the world move in an absolute space and an absolute time (Chap. 1), always remaining identical in mass and shape.
2. Newtonian mechanics is closely related to a strict determinism, i.e., the future path of a moving object can be completely predicted and its past completely disclosed if the present state is known in all details.

From the point of view of Newton's mechanics we have the following conception (vision) of the world: In the beginning God had created material objects, the interaction between them, and also the equations of motion. The whole universe was set in motion and it has continued to run ever since, like a huge machine, governed by the equations of motion. In other words, there is a giant cosmic machine that is completely causal and deterministic. A definite cause gives rise to a definite effect, and the future of this cosmos or a mechanical object can be predicted with absolute certainty, at least in principle.

The tremendous success of this mechanical picture in the eighteenth and nineteenth centuries suggested that this mechanical conception could be consistently applied to all branches of physics, that all phenomena could be explained in this way, not just the movements of planets.

This success is in particular reflected by the following fact: Due to the irregularities in the motion of the planet Uranus, it was concluded that another planet should exist. It could be predicted precisely from the calculations "when" and in "what direction of space" one should look in order to observe its appearance. Le Verrier in Paris and Adams in London made these calculations using Newton's equations of motion, and then Verrier wrote in 1846 a letter to a colleague in Berlin with the instruction to look in a certain direction at a certain time, and the colleague in Berlin actually observed the new planet Neptune. *An astronomer had discovered a new planet without raising his head*. [19] Clearly, this kind of predictions elevated the mechanistic world view, i.e., Newton's mechanics, to a gospel. In its extreme form, even living bodies should behave like a mechanical machine.

Clearly, due to the large success, the laws given by Newtonian mechanics were seen as basic laws of nature and were considered to be the ultimate theory of natural phenomena. Although we know that Newton's mechanics is a limited view, some of its basic peculiarities are still in our heads.

2.4.6 Lamettrie's View

It was Julien Offray de Lamettrie (1709–1751) who required that living human bodies (and of course animals and plants) should behave like a Newtonian machine, i.e., without any spontaneity. In Lamettrie's opinion the behaviour of human beings was also completely causal and determinate, like a planet in his movement around the sun. In other words, it was seen as being possible to predict — in principle — with absolute certainty all the future activities of a human being if his present state was known. Within this view men are merely automata and not free individuals capable of influencing the course of events by their violations. Then, we come to a completely mechanistic world in its ultimate nature, and the "mind" is a production of mechanics. Clearly, all those idealogies and perspectives about men which were based on religious considerations had no place within such a view.

Lamettrie's radical form of materialistic natural philosophy received little attention in the eighteenth and the first half of the nineteenth century. However, the concept of evolution, that is, the concept of an evolving world rather than a static one changed the situation. As we know, Charles Darwin proposed in the second half of the nineteenth century that the earth and all living things had evolved through a long history, a history of continual, gradual change (see also Sec. 2.1). The concept of an evolving world is however not in accord with certain creation myths of most religions, which have in common essentially static concepts of the world. Within such mythological pictures the world was created and has not changed since then. Darwin's evolution theory was universally accepted by serious scientists even before his death in 1882, and the following view became more and more popular: No God had created the earth and everything on it at a certain instant, but all these things had evolved throughout a long history in time in line with the laws of physics, which were assumed to be Newton's equations of motion. This view was the cause why Lamettrie's radical form of natural philosophy, completely materialistic in character, became increasingly attractive.

New physical phenomena made the limitations of Newton's mechanics apparent. In particular, electromagnetic, relativistic and quantum effects made clear that the world was not completely governed by the mechanical laws. None of their features had absolute validity. Nevertheless, the basic concept of Newton's mechanics to consider the world as a space-time container, in which all material bodies and fields are embedded, remained a basic concept.

It is therefore not surprising that even Lamettrie's vision, to consider "mind" as a production of mechanics, remained in its basic form valid. The term "mind as a production of mechanics" has merely to be replaced by "mind as a production of the physical laws", whereby the physical laws exclusively describe the states of matter. Then, we come to the modern statement "mind is created by matter".

The "container principle" is even used in modern physics. Up to the present, scientists embed the real something into space and time, and this is obviously a relic from Newton' theory. No doubt, our spontaneous impressions of the world, which appear in everyday life in front of us, suggest this container conception. The big success of Newton's mechanics has led to the stance to pass on and adopt this conception again and again, even when new theoretical ideas were envisaged. In fact, from the point of view of modern physics the vacuum (the container) has nothing to do with emptiness, just the opposite is the case: Empty space (vacuum) is a "*hyperactive player, a prolific producer of jittering fields and virtual particles. The vacuum is the most complex substance in the universe. The biggest challenge for theorists of all may simply be emptying the vacuum of all the trappings it's acquired over the past fifty years.* "*They have filled the vacuum with so much garbage, there isn't room for the cosmological constant,*" said Leon Lederman: "*Einstein freed us from the ether. Now we need to get rid of (today's version of ether) again. We need to sweep the vacuum clean.*" No doubt, the fact that the space-time is filled with a real something in modern physics is often considered as a ballast and is in the opinion of serious scientists rather unattractive.

One possible solution for this problem is to work within the "projection principle" (Chap. 1). Here no material bodies are embedded in space (space-time). Furthermore, "mind is not created by matter", as in traditional physics, but mind and its products (phantasy, etc.) can obviously exist without the material part of the brain; the source of mind is obviously not on the material level.

2.5 FINAL REMARKS AND SUMMARY

1. Cars, flowers, sun, moon and the stars cannot have the structure as they directly appear to us, that is, as they appear spontaneously before our eyes.

   How can a human being not be expected to recognize the absolute truth around him.? The answer is given by the strategy of nature, and the plan of this strategy is primary to "survive" and not to recognize the basic, objective reality. This feature is clearly reflected in the principles of evolution. Reality outside is not assessed by "true" and "untrue" but by "favourable towards life" and "hostile towards life". This strategy essentially determined the development of the recognition apparatus of man and of course also those of other biological systems.

   The inclusion of evolution is a fundamental aspect and should have in general a repercussion on the structure of the basic physical laws. If man's recognition apparatus developed according to the condition "favourable towards life" and "hostile towards life" and not by "true" and "untrue", then also the structure of the physical laws, constructed by human beings on the basis of "his" recognition apparatus, must be influenced by the same evolutionary conditions. In particular, our theoretical conception of the world outside should not lead to "absolute" statements.
2. The process, which produces the theoretical "world view" of the individual comes into existence within the brain, and this process must be consistent with the corresponding model construction. If that should not be the case, the individual could encounter problems mastering his environment and even survival is in general not guaranteed.

   No doubt, a theoretical model for the brain will be influenced by the basic structure of the laws of physics. When we want to change certain brain functions by nanotechnological means, the model for the brain has to be reliable. It is however difficult to assess the reliability of the present laws of physics if their structure is dictated by evolutionary conditions. What do we expect?

3. In Chap. 1 we gave some general statements concerning the relationship between the world outside (basic reality) and the image in front of us, which is produced by the brain. It turned out that there can be no one-to-one correspondence between the structures in these images and the structures in the basic reality. We recognized that the cause for all these facts and peculiarities is the strategy of nature (evolution).

   Man's perception of that what we call "reality" is essentially determined by the principles of evolution, in particular by the principle of usefulness. Due to that, characteristically human beings developed in accord with the following criterion: He only selects those information from the environment that is "useful" for him, which is useful for life.
4. The experienced world is essentially relevant for the development of theoretical conceptions. Therefore, these theoretical developments could be considered as complete if we were able to observe a complete world. This is however not the case. Why?

   The perception of a complete reality in the sense of a precise reproduction implies that with growing fine structures an increasing amount of information of the outside world is needed. Then, the evolution would have furnished the sense organs with the property to transmit as much information from the outside world as possible. Just the opposite is correct: The strategy of nature is to take up as little information from the reality as possible. Reality outside is not assessed by "complete" and "incomplete" but by "favourable towards life" and "hostile towards life". The perception of reality by biological systems is essentially influenced by the principle "as little outside world as possible".

   Man and other creatures are characterized by this species-preserving appropriateness. The principle of "as little outside world as possible" is compatible with the principles of evolution; it is a succession of evolution. Only those things that are useful are relevant for a human and other creatures.
5. Basic reality contains the complete information about the world. Only a part of it is projected on space and time, and this follows

directly from the principle of "as little outside world as possible". We have called the part that is projected onto space and time the "material reality". This material part is defined through our spontaneous impressions, which we have directly in front of us in everyday life, and of course by the facts we observe with suitable measuring instruments.

We are caught in space and time. This is valid in connection with observations, but also with respect to our thoughts. Therefore, the theoretical conceptions of the world, which we construct intellectually, have to be considered to be dependent on the state of evolution, which a certain biological system (for example a human being) occupies. This is almost a matter of course because we want to describe that information from the basic reality, which has been selected after the principle of "as little outside world as possible". Thus, all our theoretical views in physics only represent restricted information about the world outside, which is projected onto space and time.

6. Our "spontaneous" impressions, which we have directly in front of us in everyday life, define the "material reality", but this term also includes those facts that we observe with suitable measuring instruments. The material world consists of the bodies in front of us as well as of atoms, molecules, elementary particles and other objects that are not directly observable by means of our eyes.

   Material realities appear as "geometrical structures" that are exclusively "representations in space and time", whereby space and time are characterized by the elements x, y, z and time τ.

   This is however a construction (definition) by the human observer. In other words, we should not assume that the material entities appear as separate units in the basic reality as well. The bodies around us as well as that what we call atoms, molecules, elementary particles etc. do not reflect an "absolute" fact.

   Another type of observer, different from a human being, would observe another world, because he selects information from the basic reality which is different from ours. Therefore, the material reality of this other type of observer must be different from that of a human observer. In summary, what we call "material reality" is observer-dependent. It is a construction (definition) by the observer. The material entities do not appear as separate units in the basic reality.

There might be no information overlap between the both material realities. The bodies around a human observer as well as what he calls atoms, molecules, elementary particles etc. do in general not belong to the material reality of the other type of observer; he defines his own material reality on the basis of other entities.

Essentially, the term "material reality" does not reflect an "absolute" fact. It does not appear in the basic reality but is obviously a strict observer-dependent definition.

7. The evolutionary development of human beings and other biological systems is essentially determined by the principle of "as little outside world as possible". It is the goal of evolution to develop individuals in such a way that they are able to treat the environment optimally. In particular, the "unconscious world view" of the individual, that is the picture of the word which appears unconsciously in front of him, is fixed by this kind of selected information.

   This must however be an incomplete view of the reality outside. Due to the selection there can be no one-to-one reproduction between what we have in front of us and what is actually positioned in the reality outside (basic reality). An example for this incomplete impression of the world outside is represented in Fig. 1.

   All the "products of mind" and the "products of phantasy" etc. do not appear in such images, but have to be considered as real as the images in front of us. Both types of appearances are likewise states of the brain and reflect certain aspects of the basic reality. The intellectual, mental and spiritual features, developed by individuals, consciously create information that can be used by the individual for the improvement of life conditions.

   Both the unconscious and the conscious states have exactly the same task: They determine the possibilities to cope with the outside reality, in particular to construct favourable life conditions.

Thus, the unconscious and the conscious creation of information are equivalent in character; they are different from each other but do not differ principally. The "unconscious world view" is extended by the "products of mind" leading to an extended world view for man. In this way we obtain a "constructed reality".

Chapter Three

OTHER OBSERVERS

■ ■ ■

In this section we would like to talk about some specific questions, which concern the notions of "inside world" and of "world outside". In this regard we will point out the view of Heinz Von Förster (1911–2002). Another question is whether we may consider all observers as equivalent. Are the "pictures of reality" more or less identical for all observers? Are they independent of the biological structure? This should not be case, and we will underline this statement by specific experiments that have been done within the framework of behaviour research.

3.1 HEINZ VON FÖRSTER

Within the framework of the "projection principle" we have to conclude that the structures in the image (inside world) are different from those in the world outside. There is no one-to-one correspondence, which is in contrast to the "container theory". It is interesting to note that the famous cyberneticist and biophysicist Heinz Von Förster came to the same conclusion of the projection theory, although he used other arguments. In [20] we find the following statement:

The eminent cyberneticist Heinz Von Förster points out that the human mind does not perceive what is "there", but what it believes should be there. We are able to see because our retinas absorb light from the outside of the world and convey the signals to the brain.

The same is true of all of our sensory receptors. However, our retinas don't see colour. They are "blind", as Förster puts it, to the quality of their stimulation and are responsive only to their quantity. He states: 'This should not came as a surprise, for indeed "out there" there is no light and no colour, there are only electromagnetic waves; "out there" there is no sound and no music. There are only periodic variations of air pressure; "out there" there is no heat and no cold, there are only moving molecules with more or less mean kinetic energy, and so on. Finally, for sure, "out there" there is no pain. Since the physical nature of the stimulus — its quality — is not encoded into nervous activity, the fundamental question arises as to how does our brain conjure up tremendous variety of colourful world as we experience it any moment while awake, and sometimes in dreams while asleep. [20]

No doubt, Von Förster's reality concept is different from what is obtained by the projection theory, but we do not want to analyze this point here. It is however remarkable that, within the view of Von Förster, the structure in the inside reality is for an observer different from the structure in the outside world, and this is in accord with the projection theory, i.e., there is no one-to-one correspondence. There is a principal difference between the notions "quality" and "quantity".

For Von Förster, the basic reality is accessible; electromagnetic waves, pressure variations, etc. are embedded in the world outside (basic realty). For example, the pressure variations in the outside air produce inside the observer a certain quality that we experience inside as music, but — as Von Förster pointed out — music is not a feature of the world outside.

Clearly, the features pointed out by Von Förster have to be judged within the projection theory in another way but, as we have already remarked above, we do not want to analyze Von Förster's statements in more detail.

The fact that there is not a one-to-one correspondence between the inside world and the world outside (Chap. 1), underlined by Heinz Von Förster, is also impressively supported by specific investigations in the field of brain and behaviour research. In [21] we find a relevant comment:

Another of Lashley's discoveries was that the visual centres of the brain were also surprisingly resistant to surgical excision. Even after removing as much as 90 percent of the rat's visual cortex (the part of the brain that receives and interprets what the eye sees), he found it could still perform tasks requiring complex and visual skills. Similarly, research conducted by Pribram revealed that as much as 98 percent of a cat's optic nerves can be severed without seriously impairing its ability to perform complex and visual tasks.

Such a situation was tantamount to believing that a movie audience could still enjoy a motion picture even after 90 percent of the movie screen was missing, and his experiments presented once again a serious challange to the standard understanding of how vision works. According to the leading theory of the day, there was a one-to-one correspondence between the image the eye sees and the way that image is presented in the brain. In other words, when we look at a square, it was believed the electrical activity in our visual cortex also processes the form of a square ...

Although findings such as Lashley's seemed to deal a deathblow to this idea, Pribram was not satisfied. While he was at Yale he devised a series of experiments to resolve the matter and spent the next seven years carefully measuring the electrical activity in the brain of monkeys while they performed various visual tasks. He discovered that not only there's no such one-to-one correspondence but there wasn't a discernible pattern to the sequence in which electrodes fired. He wrote of his findings: "These experimental results are incompatible with a view that a photographic-like image becomes projected onto the cortex surface".

The results of the basic investigations by Lashley and Pribram confirm strongly the outcome of the projection principle, which is mainly based on the fact that the space (space-time) cannot be the source of physically real effects. From this follows directly that the material entities cannot be embedded in space and time; only the images, which we have in front of us, are representations in space and time. In other words, a one-to-one correspondence between the image and the world outside has to be excluded. The details have been pointed out in Chap. 1.

3.2 ESSENTIAL CONDITION: EQUIVALENCE OF ALL OBSERVERS

An observation can be considered as an objective fact when the observers are interchangeable. This is one of the fundamental principles in physics and science, respectively. The result of a measurement made by one observer has to be identical with that of any other observer. Then, the object of observation can be considered as an "objective fact" independent of the observer. Only such a kind of information can be recognized in physics, i.e., the equivalence of all human observers represents an important basic principle in science.

Not only in the laboratory but particularly in everyday life objective performances can be achieved without conscious activities. If one shows to any person a tree and asks what he sees, "a tree" will always be the answer. One would be surprised if anyone would see a car or a house instead of a tree. This is more like a matter of course, but it is not trivial.

In Sec. 2.2.3 we studied a human observer, marked by S, and an observer of another type that we have marked by S'. The material reality of S' was assumed to be different from the material reality of the human observer S. Both S and S' select spontaneously from basic reality (having the information content C) a certain part; it is A in the case of S and A' in the case of S'. Due to the principle of "as little outside world as possible" we have $A < C$ and $A' < C$. Furthermore, we have $A' \neq A$ since we assumed that S' is different from S. In other words, the information A' should be different from information A. Information A defines the material reality of S, and A' defines the material reality of S'. What we call the "material reality" is observer-dependent. It is a construction (definition) by the observers, in this case S and S'.

Within the projection theory the material reality appears as a "geometrical structure" within the brain of an observer. Since both observers, S and S', are assumed to be different from each other, the frames on which the facts are projected should be different from each other. In the case of S we have the space-time elements x, y, z and time τ, and in the case of S' we have marked the elements of the projection frame by a, b, c, ... (Sec. 2.2.3).

How a human observer S experiences the world in front of him is known (cars, trees, houses, etc.), but we can at first say nothing about how the observer of another kind (marked by S') experiences "his" world; it remains hidden to S. Sure, we can possibly picture the observer S' within the space and time of S, i.e., on the basis of the elements x, y, z and τ, and S' can possibly picture S on "his" projection frame with the elements a, b, c, ..., but the spontaneous impressions in front of S', which appear in "his" everyday life, are different from the familiar impressions in front of S, which — in other words — the human observer experiences spontaneously in "his" everyday life.

This is at first a more or less abstract construction. However, we know that there are biological systems that are different from each other, and we cannot assume that the spontaneous events and things, which appear spontaneously in front of them, are similar or even identical to ours. We often assume that, but this must not be the case. Already the principle of usefulness excludes that. It is interesting to note that relevant results obtained within the frame of behaviour research strongly support this view. Let us discuss this point in somewhat more detail.

3.3 OTHER BIOLOGICAL SYSTEMS

In [23] and in the sections above, we came to the conclusion that the picture of reality must be species-dependent. In other words, we have to conclude that the actions of other biological systems are in general based on a picture of reality that is different from that of the human observer. How can we verify that? Wolfgang Schleidt performed some interesting experiments using a turkey, its chick and a weasel which is the turkey's deadly enemy, and he studied the behaviour of the turkey in order to learn something about the perception apparatus of the turkey. Let us repeat here the main facts:

Schleidt worked with more or less everyday methods. However, his experiments could be of such importance as certain key experiments in physics which have fundamentally changed the scientific world view. Schleidt demonstrated convincingly that the perception apparatus of a turkey must be quite different from that of humans. These experiments led to dramatic and unexpected results and demonstrate that the turkey must experience the world optically quite differently from the way we do, even though the eyes of the turkey are quite similar to ours. There is obviously no similarity between what the turkey experiences and what a human being sees in the same situation.

This can be well explained within the framework of the projection theory. The spontaneous impressions in front of the turkey (here *S*'), which appear in its everyday life, are obviously different from the familiar impressions in front of a human observer *S*, which he experiences spontaneously in "his" everyday life.

Both systems, man and turkey, react correctly in the normal case because both species are able to exist in the world, which can only be possible from the point of view of the modern principles of evolution if their particular views of the world are correct. Therefore, although the conceptions of the world of man and turkey are on the one hand different from each other, they are on the other hand correct in each case. This means that neither of these two conceptions of the world can be true in the sense that they are a faithful reproduction of nature: Objective reality (basic reality) must be different from the images which biological systems construct from it. We already came to this conclusion on the basis of space-time arguments. A detailed description of these experiments together with conclusions and interpretations is given in [23].

The results for the turkey can be generalized because there is no reason to believe that turkeys have to be considered as an exception. Without doubt Schleid's results support strongly our view concerning the notion of "reality" developed above.

In summary, the experiments by Schleidt deliver essential contributions to our understanding of that what we call reality and they can help us to learn something about the relationship between reality and any kind of observer (man, turkey, etc.). However, we can only say that the perception apparatus of the turkey is different from that of humans; the details are not accessible in any way, but we do need more information for answering such principal questions. We even do not know at first if such biological systems (turkeys) experience "their world" within the framework of space and time. However, when we identify the turkey with observer S', we have actually to assume that not only the selected information of S' from the basic reality is different from that of S (human observer), but also the frame on which this information is projected. The human information is projected onto a frame with the elements x, y, z and τ, and the turkey's selected information might be projected on another frame having the elements $a, b, c, \ldots$, where the elements $a, b, c, \ldots$ are unknown.

Sure, we can picture S' (turkey) within the space and time of S, i.e., on the basis of the elements x, y, z and τ, and the turkey S' can possibly picture a human S on "his" projection frame with the elements $a, b, c, \ldots$, but the spontaneous impressions in front of S', which appear in "his" everyday life, are different from the familiar impressions in front of S, which the human observer experiences spontaneously in "his" everyday life.

The important and basic results of Schleidt's experiments with the turkey, a guide for science, lead to the conclusion that the turkey must optically experience the world quite differently than a human does, although the eyes of the turkey are quite similar to ours. There is obviously no similarity between what the turkey sees and what a human being experiences in the same situation.

Again, the eyes of the turkey (observer S') are similar to the eyes of a human observer, but the optical impressions, which the turkey has spontaneously in front of it, are obviously not similar to the spontaneous impressions in front of the human observer; Schleidt's experiments do not allow any other interpretation. The model for this situation, which seems to contain a contradiction, is simple. The eye of observer S' (turkey), consisting of a lens and the retina, must have within the turkey's frame (with the elements $a, b, c, \ldots$) another geometrical shape than that within the frame of S (human observer), which is characterized by the elements x, y, z and τ. In a nutshell, the eye of the turkey within its own projection frame $a, b, c, \ldots$ is different, possibly quite different, from the turkey's eye within the projection frame of a human, having the elements x, y, z and τ.

The cause for the similarity of the turkey's eye and the eye of a human probably means that the "mechanisms" to form a picture are similar for both the turkey and a human.

The mechanism for the construction of a picture within the head of an observer (S or S') is given by geometrical optics and certain brain functions. Then, "similar mechanisms" means that the "principles of geometrical optics" within the projection frame (S'-space with $a, b, c, \ldots$) of the turkey S' are similar to those within the projection frame (S-space with x, y, z and τ) of the human being S. This model is suggested when we try to assess the peculiarities of a turkey from the point of view of a human being, which is exclusively based on S-space. Then, for the modeling of the turkey's view of the world, we have only to replace the information A and the projection frame of x, y, z, τ (S-space) by the corresponding features of the turkey (A', $a, b, c, \ldots$ (S'-space)); more details are given above, in particular in Sec. 2.2.3.

However, a mixing of peculiarities is not allowed: Situations in connection with the turkey may not be described on the basis of x, y, z, τ; on the other hand, situations with respect to the human being may not be described on the basis of $a, b, c, \ldots$.

The cause for that is obvious: The elements within the projected images of one species are in general not compatible with those of another species; the elements of pictures of different species are in general not interchangeable. If one tries it nevertheless, problems can appear as in the case of Schleidt's experiments. He tried to project the elements of pictures, which are characteristic to human behavior, onto the projection frame of the turkey and the result was a terrible mess [22].

For comparison, it is not possible to drive an electric car with petrol because petrol is not compatible with an electric engine. On the other hand, it is not possible to drive a car with petrol engine with the battery of an electric car. Petrol is not useful for the electric car and a battery is not appropriate for the petrol engine.

We judge the picture of reality, which the turkey has in front of it, on the basis of S-space, and this is wrong. The turkey's picture of reality is based on S'-space $(a, b, c, \ldots)$ and not on S-space (x, y, z, τ). The assumption that the turkey's "world" is based on S-space, the world of the human being, led to a disaster as Schleidt's experiments demonstrated.

3.4 GEOMETRICAL OPTICS

A human observer (let us mark him again by the letter S) has a certain image of his environment in front of his eyes; it is an image like what is given by Fig. 1. This image is a representation in the "space of S". Let us assume that this space-time image contains another human observer, say S_A, and a tree. The image in front of observer S_A, represented in the "space of S_A", is almost the same, which S experiences; it is not exactly the same because the position of S_A in the "space of S" is different from the position of S. However, since both human observers are equivalent, they have images in front of them with exactly the same features. In particular, the "space of S" is identical with the "space of S_A".

Observer S_A appears in the "space of S" in a form we are all familiar with: Observer S_A has a head, eyes, legs, etc. In particular, observer S recognizes that each eye of observer S_A has a lens. How can S judge the role of the eyes of observer S_A? Quite generally, we may state the following: The objects, positioned in the space of S (in the picture of S) can be manipulated by optical instruments. When we use a certain kind of optical lens we can magnify the objects in the picture.

What does that mean within the framework of the projection theory? We construct, within the "space of S", the magnified object (tree) by means of the lens and the tree which we experience without the lens. The one occupies space on the left-hand side of the lens, the other, the magnified tree within observer S_A, occupies space on the right-hand side of the lens, where the retina is. In this way observer S constructs an object, in this case the tree, within the brain of observer S_A. That is, both trees are in the same space, as it is in our example the "space of S".

What happens with the magnified tree on the retina of observer S_A? In order to answer this question we have only to repeat what is pointed out in Appendix A: Observer S_A does not register the picture of the tree on the retina inside the eye, but he has the impression that he is standing opposite the tree which is located in the space outside, not standing on its head but upright. Also S_A sees "real objects" in front of him and around him, which are positioned in his space, that is, within the "space of S_A". Within this act of perception, the eye, the optic nerves and the brain of S_A work together. To see without the brain is as impossible as to see without eyes.

Clearly, both trees that observer S recognizes cannot be outside of S because there is no space outside. Such lens-constructions only allow one to say something about the objects in the picture of S (within the "space of S"), without lens and with lens; nothing can be said about the object and the situation in the reality outside. The only thing we can talk about is the relative changes of the objects (trees) in the picture of reality, nothing else. The investigation of the relationship between the object outside and the object inside of S is not possible.

If we however replace the human observer S_A in the picture of S by a turkey, which is not equivalent to a human observer, such a construction is not possible. The reason is obvious: The situation, judged from the point of view of observer S, is not compatible with the image in front of the turkey. In particular, the "space of S" is identical with the "space of S_A", but is not identical with the space of the turkey. We inevitable come to a wrong assessment. The experiments by Wolfgang Schleidt support that impressively.

The following example is instructive (let us repeat here the text given in [23]): A certain fact can be expressed linguistically in many but equivalent ways. For example, the fact "In the beginning was the Word", formulated in English, cannot be recognized if we express for example the same fact in Russian. The symbols of the Russian language are quite different from those of the English language. Not only the symbols, but also the rules after which the symbols are connected (grammar), are different from each other. Each language, i.e., English or Russian, forms a closed consistent system. However, a

Russian cannot understand an English sentence if he does not speak English and, on the other hand, an Englishman cannot understand the Russian sentence if he does not speak Russian.

Exactly the same should be true for man and turkey. Both form spontaneously pictures in front of them, which are obviously quite different from each other. The turkey does not understand the image in front of a human being and, on the other hand, a human being does not understand the image, which is in front of a turkey. As we have stated in Chap. 2, the image in front of a species does not have to be true in the sense of a one-to-one reproduction, but it must merely be useful, which follows from the principles of evolution (see in particular Chap. 2). We remarked several times that no biological system has access to the basic reality, and only the basic reality contains the absolute truth.

3.5 DIFFERENT IMAGES OF THE SAME OBJECT

There are only a few things of which the scientist is more convinced than of the real character of his experiments and the objectivity of his statements about them [24]. This statement describes the present situation correctly. Here two notions are in the focus, "reality" and "objectivity". Most scientists believe that they have the absolute truth in their hands; they are in particular convinced to have "basic reality" in front of their eyes, and they are also firmly convinced that all these facts reflect an absolute objectivity.

All that becomes questionable when we work within the "projection theory"; here nothing observable exists independent of an observer. In contrast to the statement above, within the projection theory a human observer can nothing say about the basic reality; it exists but its structure remains principally hidden. Our cognition apparatus can only form pictures of reality as, for example, a chick used in the experiments of Wolfgang Schleidt.

The same holds for other biological systems: The cognition apparatus of a turkey also forms pictures of reality, which are however different from those formed by a human observer. The chick seen by a turkey is different from the same chick seen by a human observer. In other words, different biological systems form in general different images of the same object of basic reality.

Since we are not able to say something about the basic reality, i.e., the absolute truth, no human being is able to formulate what is often called the “world equation”. Clearly, a human can formulate a specific “world equation” for his restricted world view, but a turkey could in principle also formulate a specific “world equation” using the facts of its specific world view. However, both world equations would be different from each other because they are based on different facts, which are not compatible.

The images we have in front of us do not have to be true in the sense of a precise reproduction, but have “only” to be conclusive and their elements must be compatible. As we have remarked above, petrol is not appropriate to the electric car, a battery is not compatible with an engine working on petrol.

Furthermore, only those facts which are experienced by observers of the same species can be considered as “objective”. Since a turkey and a human being are not of the same kind, we may state that the two objectivities, that of man and that of a turkey, are not identical. “Objectivity” does not mean that a certain fact actually exists in the basic reality in the form experienced by the biological systems, here man and turkey.

3.6 CONSTANCY PHENOMENA

The basic reality is not accessible to human beings. But we have “levels of reality”, which reflect certain features of the basic reality. In Sec. 1.11.5 we introduced the principle of level-analysis, and we explained this principle with the help of only two levels L_1 and L_2 (see Fig. 17; a more detailed discussion is given in [2]): All levels of reality reflect certain peculiarities of the same world outside (basic reality). Level L_1 represents the “material level”, and on level L_2 the theoretical conceptions of the world are positioned. Whereas the “objects” of level L_1 are geometrical entities embedded in space and time, the theoretical conceptions cannot be described in space and time. The "objects" of level L_2 are the physical formulas. Because a theoretical formula is more general than a certain entity (material object) in space and time, level L_2 is positioned in Fig. 17 above level L_1. Once again, both levels are constructions by the observer and belong to the brain. All that has been pointed out in Secs. 1.11.3–1.11.5.

3.6.1 Connections to the Basic Reality

Again, each level of reality reflects certain peculiarities of the same world outside (basic reality), but where do we know the "objects" on the various levels? There must be a certain connection between the observer and the basic reality. What are the objects of the projection theory?

Due to the principle of "as little outside world as possible" the observer selects a certain information A from the total information C, which the basic reality contains. However, we do not know the selection process because we know nothing about information C, and we also know nothing about the transformation process, which organizes the projection of the selected information A onto space and time.

These space-time structures are the "basic information" and come into existence spontaneously in an unconscious way, i.e., without thinking. We have these images in front of us without to know how they came about. This is the case for the material level L_1 on which the information A is projected. Clearly, we extend this information consciously through measurements, but the basic image that we have in front of us in everyday life comes into existence spontaneously without any conscious action by the human observer. This is obviously the most direct connection that we have to the basic reality.

How did the "objects" of level L_2 come into existence? Where does the observer look to find the information T, which belong to level L_2? In other words, from where can the observer know the physical formulas? Answer: A human observer experiences them by "intuition"; here intuition means the direct perception of the truth independent of any reasoning process, i.e., the perception of the basic truth on level L_2

takes place unconsciously. It is simply an immediate comprehension, and the observer cannot say how or why. The basic information T of level L_2 comes into existence spontaneously; note that the information A of level L_1 also comes into existence spontaneously.

As we have pointed out above, level L_2 reflects certain peculiarities of the same world outside (basic reality). Thus, the grasp of a certain fact by intuition means that there must be a certain connection between the observer and the basic reality, but we are not in the position to say how this connection works.

3.6.2 Conscious Objectivation

An idea comes into existence spontaneously without thinking, but the unconsciously grasped idea is not pictured in space and time as in the case of the material objects positioned on level L_1. However, the idea itself has for the present not much meaning; an idea becomes first useful for a physicist when the essence of it is clothed into mathematical expressions, and this is a "process of thinking". That is, a physical formula, an object of level L_2, comes finally into existence in a conscious way. This process, the use of an idea for the development of a mathematical formula and its scrutiny, can be called "conscious objectivation" or "objectivation by thinking". Let us briefly explain why.

Objectivation by thinking here means that a physical view of the world, expressed by mathematical expressions, is tested via a dialogue with nature (Sec. 1.6). In this way a world view can be improved and, if necessary, also be rejected. The theory should be reflected in many, if possible in all real physical situations, i.e., a theoretical conception should be in accord with many experimental findings. This has to be verified, and defines a certain process, which can be called the "process of objectivation". Such a procedure is meant when we use the term "objectivation by thinking", which in particular means "conscious objectivation". In other words, within the projection theory a certain world view is "discovered" by objectivation. A world view that is useful should be constantly reflected in many variations of thinking and different experimental configurations.

Karl Popper made clear that scientific laws as, for example, Newton's equations of motion, cannot simply be the result of direct experience or measurement either. He said (see also Sec. 1.9.3):

Groping in all directions. I do not favour the picture of science as gathering observations and distilling the laws from them. This conception is utterly wrong. It mechanizes the creative act of human thinking and inventing. To make that point clear is most important to me. Science proceeds in a different way, that is, it proceeds in such a way as to test ideas and world views.

The term "groping in all directions" means to test a physical view of the world via a dialogue with nature (Sec. 1.6). In this way a world view can be checked and possibly improved or rejected. The theory should be in accord with the results obtained from many experimental configurations, the more the better. That has to be verified. Such a process can be called "objectivation by thinking". Once again, a reliable theory and world view, respectively, should be constantly reflected in many variations of thinking and different experimental configurations. We "discover" the theoretical physical laws.

3.6.3 Unconscious Objectivation

What about the "material entities" on level L_1, which appear spontaneously just in front the observer as geometrical structures in space and time? Konrad Lorenz (1903–1991) convincingly showed that these material entities are also discovered by objectivitation, just like the scientific laws. This process reflects in the case of material objects certain "unconscious actions" on the basis of physiological mechanisms, which are known as "constancy phenomena". What does this notion mean?

Objectivation in an unconscious way is for a human being a relevant factor. The recognition of certain situations in the environment requires reliability, i.e., the occurrences must be reproducible. How has evolution realized that? It caused organisms to develop physicological mechanisms which register automatically and objectively those signals from the environment that are relevant and useful for the biological system, and only a part of the information from the basic reality is selected. In this regard, it is most important that an object, observed by the individual, is unambiguously recognizable even in the case of large variations in the environment. In order to guarantee this, evolution has developed "constancy mechanisms" where their objectivation performance is managed by a complex physiological apparatus. Konrad Lorenz wrote [25]:

Of special interest to the scientist striving for objectivation is the study of those perceptual functions which convey to us the experience of qualities constantly inherent in certain things in our environment. If, of course, we perceive a certain object (say a sheet of paper) as "white", even when different coloured lights, reflecting different wavelengths, are thrown on it; this so-called constancy phenomenon is achieved by the function of a highly complex physiological apparatus which computes, from the colour of the illumination and the colour reflected, the object's constantly inherent property which we call its colour.

Other neural mechanisms enable us to see that an object which we observe from various sides retains one and the same shape even though the image on our retina assumes a great variety of forms. Other mechanisms make it possible for us to apprehend that an object we observe from various distances remains the same size, although the size of the retina image decreases with distance.

Constancy phenomena thus allow objectivation at the material level in an unconscious, non-intellectual way by the physiological apparatus. Those processes and objects that are relevant for survival appear directly and reliably in front of a human being, i.e., without conscious action of the individual. This makes it seem to us that there is an independent reality, independent of the individual. This is because of the constancy of the phenomena which allows that the experienced reality becomes concrete in character. Furthermore, this reality appears as a complete world since all unnecessary information about the environment has been eliminated. The following remarks by Konrad Lorenz are instructive [25]:

What causes us to believe in the reality of things is in the last analysis the constancy with which external impressions recur in our experience, always simultaneously and always in the same pattern, irrespective of variations in external conditions or in our physiological disposition.

At this point we must emphasize once again that it is not of primary importance to have a true image of the world in front of us in the sense of a precise reproduction. The images are produced for the purpose to help the individual to cope with life successfully. The registration of bodies (and processes) without conscious action, as well as their reliable identification, makes them appear independent and concrete in character. These material entities appear as geometrical structures in space and time and are positioned on level L_1 in Fig. 17.

The immediate execution and visualisation which is essential for survival at least in the phylogenetically early phase demand such concrete relations. However, as already pointed out, such concrete images on the material level are only incomplete representations of the real world outside. At least in the phylogenetically early phases all those things were ignored which were not of relevance for the naked survival.

The capability of thinking was developed during the later phase of evolution. The process of thinking makes it possible to extend the individual's knowledge about the world. Relationships can be constructed and analyzed in a theoretical way and this new knowledge can be used to improve the conditions of life.

With the occurrence of mind, individual experience became possible, which was not possible before. Individuals without a conscious mind handle certain situations in their life not on the basis of their own experiences, but through behavioural patterns which have been developed by the members of thousands of generations of their own species. Such behavioural patterns lose their value immediately, or even become harmful, when the environment has changed crucially. There are a lot of examples for this in nature; species have become extinct due to changes in the conditions of the environment. By the process of thinking, we can adapt our behaviour due to our own experiences; the individual can now react to quick changes in the environment.

Here thinking means objectivation in a conscious, intellectual way, which we have already discussed above in connection with the physical laws (Sec. 3.6.2). We just recognized that at the level of material objects objectivation processes take place too. It is however an objectivation in an unconscious, non-intellectual way by physiological processes, which we have discussed in connection with Konrad Lorenz's remarks.

In essence, the material entities as well as the scientific laws (world views) are registered (observed) in the same manner, i.e., by objectivation processes. We may therefore consider conscious (intellectual) and unconscious (non-intellectual) observations of facts as analogous processes. Konrad Lorenz remarks [25]:

The physiological functions underlying these constancy phenomena are of greatest interest in the context of theory of knowledge because they are exactly parallel to the process of deliberate, rational objectivation referred to above.

On the basis of scientific law, a lot of specific theoretical configurations can be constructed (selected) consciously by thinking, and these theoretical configurations have to be compared with suitable experimental results. If the various theoretical images agree with the corresponding experimental findings, we may state that the scientific law is constantly reflected in many different experimental configurations. The processes of thinking, which carry out such selections, can also be regarded as "constancy phenomena".

3.6.4 Conclusion and Final Remarks

The "things" at the various levels represent states of the brain and reflect certain features of the basic (objective) reality. What are discoveries and what features are inventions? We do not know the details about the mechanisms within the observer's brain. However, we may state quite generally that the observer picks up certain information A from the basic reality, i.e., he discovers it, and this information is projected as an image onto space and time (level L_1, Fig. 17). However, we know nothing about the selection process and the transformation of this information from the basic reality into the observer's brain. Space and time do not exist in the basic reality (Chap. 1).

That information, which is not describable in space and time, is given symbolically in the form of equations (level L_2, Fig. 17). Thus, the basic information is a discovery whereas the pictures and the equations should be considered as an invention of the observer. These discoveries are done on the basis of the so-called objectivation processes, and we have to distinguish between conscious and unconscious objectivations.

Conscious actions

We develop a mathematical equation, i.e., a theory or a world view, by thinking on the basis of a discovered idea. The theory (equation) should be in accord with the results obtained from many experimental configurations, the more the better. For this purpose the general theory has to be adapted to real, specific situations, i.e., each experimental configuration needs specific forms (models) of the general theory. The agreement of the results obtained from the theoretical models and the experiments has to be verified. In this way the general mathematical equation (theory) can be objectified and can be considered as real. Such a process can be called "objectivation by thinking". Once again, a reliable theory and world view, respectively, should be constantly reflected in the many variations of thinking (models) and different experimental configurations. We "discover" the theoretical physical laws!

A physical law is a "product of mind". The products of imagination (phantasy) belong to the class of mind products. Thus, similar criteria — perhaps to a slight extent — should be fulfilled in connection with the products of imagination: A literary picture very often also sums up reality in a single "image" (here in the form of a metaphor). This, however, can be applied in many situations of life, and the same "image" (metaphor) is reflected in the experience of many people. We may therefore state that the author of a narration found this "general image" by a process of objectivation, just as in the case of a physical theory. More details are given in [26] and in Sec. 1.14.2.

Unconscious actions

The objectivation processes in connection with the material objects are of particular interest (see also [26] and the literature therein). A material object observed by the individual is unambiguously recognizable, even in the case of large variations in the environment. This is guaranteed by so-called constancy phenomena where their objectivation performance is managed by a complex physiological apparatus. Constancy phenomena thus allow objectivation in an unconscious, non-intellectual way by the physiological apparatus.

Also the objects recorded by measuring instruments are also observed unconsciously. At the level of the measuring instruments the observer imposes restrictive conditions on nature. This is done by the development of specific measuring instruments, leading to construction conditions that are imposed by the observer in order to get a specific answer on a specific question. Due to these specific construction we register only a few specific signals from a multitude of possible events in nature. The selection or construction of a measuring instrument has, in a certain sense, the meaning of a constancy mechanism, which we have discussed for unconscious objectivations, where the objectifying performance comes into existence through a complex physiological apparatus. The registration (observation) of signals in connection with measuring instruments also takes place automatically, i.e., they are recorded without our conscious help, although the selection of a measuring instrument is a conscious act.

A common principle

We have stated above that objects belonging to the various levels of reality are equally real, and we observe all these objects at the various levels by a common principle, which can be called the "principle of objectivation". This principle is valid at each level, i.e., it can be applied independent of which level an object is located. We distinguish between "objectivation by thinking" and "objectivation in an unconscious, non-intellectual way". The principle of objectivation supports the view that there is no principal difference between the "states of mind" and the "states of matter"; there are only gradual differences. Both objectifying procedures should be classified as observation processes since the basic information comes from the basic reality.

Levels of observation

The differences between the objects on the various levels is not only reflected in their theoretical description but also in their effect on us. The kind of objectivation is of special importance (Sec. 1.11.7). In principle, the objectivation processes on the various levels are different from each other. Since the basic information have to be considered as discoveries and not as inventions, we may state that each "level of reality" is accompanied with a "level of observation". Because the features of the objects on the various levels are defined differently, the methods of observation must vary from level to level. (Sec. 1.11.7).

From the process of objectivation emerges the "objects", and these produce certain level-specific feelings inside the observer. We feel the effect of objects, which are objectified unconsciously (hard objects like trees, cars, etc.), differently from those that are objectified consciously (scientific laws, products of imagination, etc.). In principle, many objectivation methods should be possible depending on the number of levels and their features.

We concluded that unconscious and conscious objectifying are analogous processes. They both select certain information from a diversity of objects and processes. Both objectifying procedures should be classified as observation processes.

Unconscious objectifying takes place on the level of the five senses (level L_1, which can be called the "macroscopic level") and also in part by measuring processes, even when here the human consciousness is essentially involved.

Conscious objectifying reflects an observation process through thinking and takes place at another level of reality (level L_2). Here the basic information about the "objects" are also selected from the basic reality. This process, however, is close to the one which takes place on the level of the five senses and measuring instruments. Konrad Lorenz has already pointed out the strict analogy between unconscious objectifying through the physiological apparatus and conscious, intellectual objectifying (see also [25] and Sec. 3.6.3). Thinking is therefore a certain kind of observation procedure, and in fact a very versatile one. It is not new to state that one can think on different levels.

The common features of level L_1 (material objects) and of level L_2 (physical laws) can be summarized as follows:

1. Both the material objects as well as the physical formulas are likewise positioned in the observer's brain.
2. The basic information of material objects and physical laws are discoveries.
3. The representation of the basic information have to be considered likewise as invention.
4. Both types of entities come into existence spontaneously.
5. In both cases (material objects and theoretical laws) certain features of the basic reality are observed, but a direct statement about the basic reality is not possible.

These points underline impressively that material objects and the physical laws are similar in character. There is a difference between them and, therefore, they are positioned on different levels, and this is because they express certain features of the basic reality in terms of more general statements (Sec. 1.11.5), but there is no principal difference between them.

3.7 SUMMARY AND FINAL COMMENTS

1. The results of the investigations by Lashley and Pribram confirm strongly the outcome of projection principle: A one-to-one correspondence between the image and the world outside has to be excluded.

Projection theory is mainly based on the fact that the space (space-time) cannot be the source of physically real effects. In particular, material entities cannot be embedded in space and time. Only the images, which we have in front of us, are representations in space and time. Lashley and Pribham worked within the frame of the brain and behaviour research, but they came to the same result: There is no one-to-one correspondence between the image and the world outside has to be excluded. The thoughts by Heinz Von Förster underline all these findings.

2. We studied a human observer, marked by *S*, and an observer of another type marked by *S*'. In general we have to assume that the material reality of *S*' is different from the material reality of the human observer *S*. Within the projection theory the material reality appears as a "geometrical structure" within the brain of an observer. Since both observers, *S* and *S*', are assumed to be different from each other, the frames on which the facts are projected should be different from each other. In the case of *S* we have the space-time elements x, y, z and time τ, and in the case of *S*' we marked the elements of the projection frame by $a, b, c, \ldots$ (Sec. 2.2.3).

   How a human observer *S* experiences the world in front of him is known (cars, trees, houses, etc.), but we can at first say nothing about how the observer of another kind (marked by *S*') experiences "his" world; it remains hidden to *S*.
3. The picture of reality must be species-dependent. In other words, we have to conclude that the actions of other biological systems are in general based on a picture of reality that is different from that of the human observer. Wolfgang Schleidt performed some interesting experiments using a turkey, its chick and a weasel which is the turkey's deadly enemy, and he studied the behaviour of the turkey in order to learn something about the perception apparatus of the turkey. Result: A human observer obviously experiences the world in front of him quite differently from that of a turkey. This surprising result agrees with the basic statements of the projection theory.

4. Due to the principle of "as little outside world as possible" the observer selects a certain information A from the total information C, which the basic reality contains. However, we do not know the selection process because we know nothing about information C, and we also know nothing about the transformation process, which organizes the projection of the selected information A onto space and time.

   These space-time structures are the "basic information" and comes into existence spontaneously in an unconscious way, i.e., without thinking. We have these images in front of us without to know how they came about. This is the case for the material level L_1 on which the information A is projected.

   How did the "objects" of level L_2 come into existence? Where does the observer look to find the information T, which belong to level L_2? In other words, from where can the observer know the scientific ideas which lead to physical formulas? We have also answered this question: A human observer experiences them by "intuition". Intuition means the direct perception of the truth independent of any reasoning process, i.e., the perception of the basic truth on level L_2 takes place unconsciously, and the observer cannot say how or why. The basic information T at level L_2 comes into existence spontaneously as in the case of level L_1 where the material bodies are positioned. As we have pointed out above, level L_2 reflects certain peculiarities of the same world outside (basic reality). However, from T we develop the mathematical formulas (physical laws) in a conscious way.
5. The objects of both levels are observations, i.e., not only material entities are observed but the physical laws as well. Physical laws are observed by conscious objectivation and, on the other hand, material entities by unconscious objectivation.
6. The material entities as well as the scientific laws (world views) are registered (observed) in the same manner, i.e., by objectivation processes. We may therefore consider conscious (intellectual) and unconscious (non-intellectual) observations of facts as analogous

processes. Constancy phenomena allow objectivation in an unconscious way by the physiological apparatus.

7. The objects (material entities, physical laws) belonging to the various levels of reality are equally real, and we observe all these objects at the various levels by a common principle, which can be called the "principle of objectivation". This principle is valid at each level and supports the view that there is no principal difference between the "states of mind" and the "states of matter"; there are only gradual differences. Both objectifying procedures should be classified as observation processes since the basic information comes from the basic reality. The unconscious and conscious objectifying procedures select certain information from a diversity of objects and processes.

Chapter Four

SPECIFIC SPACE-TIME PHENOMENA

■ ■ ■

In Chap. 1 we have justified in detail that space and time may not be the source of physically real effects. Concerning this point we have a problem in conventional physics; the effect of inertia comes into existence through an interaction between the material bodies and space. In this chapter we will deepen the situation in conventional physics where the material bodies are embedded in space (space-time), i.e., we work here within the frame of the "container principle". Newton's mechanics as well as the Theory of Relativity are of particular interest when we talk about the effect of inertia, just in connection with Mach's principle.

However, the effect of inertia must also be discussed within the frame of the conventional quantum theory. What happens with such inertial effects when we go from the classical theory to the description of (conventional) quantum effects for which Schrödinger's equation is responsible. This point is relevant for the understanding of such phenomena and should not be underestimated.

The physical conception of the projection theory is quite different from that of conventional physics, i.e., when we go from the "container principle" to the "projection principle" where the material reality is not embedded in space and time but is projected onto the space-time frame. Here the following question can be answered: Is the effect of inertia really a fundamental notion? The answer is no. We will analyze the situation in this brief chapter.

4.1 NEWTON'S MECHANICS AND THEORY OF RELATIVITY

In Newton's mechanics the notion of "inertia" is of particular relevance. As we have already remarked several times, here all real bodies are embedded in space. Even when a body does not interact with other systems, the effect of inertia is in effect, i.e., also in this case the body moves through space with a constant velocity $\mathbf{v}$:

$$\mathbf{v} = const, \tag{9}$$

where the velocity of zero is included. Equation (9) reflects the law of inertia. Within Newton's theory the effect of inertia is entirely due to the interaction of the body with space. Equation (9) is also valid for a lone body positioned in space.

The body moves relative to space, and its $\mathbf{p}$, E-state is dependent on its velocity $\mathbf{v}$ relative to space, i.e., the variation of $\mathbf{v}$ leads to changes of the body's $\mathbf{p}$, E-state. The momentum is given by $\mathbf{p} = m_0\mathbf{v}$ and the energy by $E = m_0\mathbf{v}^2/2$.

Newton's mechanics is based on an "absolute space" and also on an "absolute time". However, such conceptions are not acceptable because they reflect un-physical elements. We have analysed this point in Chap. 1 and in Appendix B.

Space and time should not be the source of physically real effects but this is the case in connection with Newton's inertia. In fact, we can never observe such a space (space-time) because its elements (the coordinates x, y, z and the time τ) are in principle not accessible to empirical tests. In other words, coordinates x, y, z and time τ can never be observed in isolated form. Also isolated space-distances between two space points and isolated time-intervals are not accessible to empirical tests. We can only say something about distances in connection with bodies and time intervals on the basis of physical processes. In other words, an "empty" space-time as a physical-theoretical conception should not be existent. This is a fundamental point and may not be neglected in the basic formulation of any physical law [1].

This was the reason why Ernst Mach required an essential principle: A body does not move in un-accelerated or in accelerated motion relative to space, but relative to the centre of all the other bodies (masses) in the universe. This requirement is often discussed in literature under the notion of "Mach's Principle". However, Mach's principle is not fulfilled in Newton's theory.

In contrast to Newton's theory, space and time of the Special Theory of Relativity are no longer independent of each other; they are tied together into a space-time. Is Mach's principle fulfilled within the Special Theory of Relativity? Definitely not! Newton's three-dimension space is merely extended here to a four-dimensional space-time, without overcoming the absoluteness. In other words, instead of Newton's three-dimensional absolute space we have a four-dimensional absolute space-time within the Special Theory of Relativity. Also this space is — as in the case of Newton's three-dimensional space — the seat of the absolute forces of inertia. Clearly, Mach's principle is not fulfilled within the Special Theory of Relativity. Also the four-dimensional space-time is the source of physically real effects (inertia).

In order to relativize the forces of inertia, Einstein was led to formulate the General Theory of Relativity. In this way he wanted to eliminate the absolute space-time. In regard to Mach, Einstein argued that the inertia of a body cannot come into play through a space-time effect but should be completely due the other matter in the universe. This relativity of inertia was the foundation of his entire considerations.

With the formulation of the General Theory of Relativity Einstein took up a completely new direction. This theory represents a magnificent structure, and its results are confirmed by many experiments. However, the General Theory of Relativity failed to fulfil its initial goal, namely to eliminate the absoluteness of space-time of the Special Theory of Relativity. Also here the space-time can be the source of physically real effects (inertia). The space-time situation within the General Theory of Relativity is outlined in Appendix B. Let us repeat here only one point:

Within Newton's theory a lone body may move through space with constant velocity indicating the effect of inertia (see Fig. 5b). As we have pointed out in Sec. 1.3.4, exactly the same effect is possible within the General Theory of Relativity. Willem de Sitter demonstrated in the year 1917 that Einstein's field equations lead to the effect of inertia in the case of a lone body moving through space-time, i.e., there is exactly that type of inertial motion which is defined within Newton's mechanics. In a nutshell, *the absoluteness of space, which Newton has claimed, and which Einstein may have attempted to eliminate, is still contained in Einstein's theory* [6].

What could be the source of inertia within the General Theory of Relativity? Why do bodies perform an inertial motion? As we have remarked, to explain this physically Ernst Mach supposed that the acceleration of a body relative to the stars has to be considered. With today's knowledge we would have to take into account, instead of merely the stars, the large-scale distribution of the masses in the universe. With regards to this idea gravitational fields would always exist as long as any single masses exist. An inertial field would appear when the body is accelerating relative to the entire matter in the cosmos.

Could this concept really be the source of inertia? How should such a mechanism work? If a train stops suddenly, am I thrown forward by the masses in the entire universe? A simple causal connection cannot be recognized. The train conductor suddenly brakes because a signal has gone on "stop". In other words, this is not caused by the cosmic masses, but by the automatically working signal box due to the information relayed about other trains. Information relayed from the galaxies to us would take thousands or even millions of years, and this is because of the long path and the finite propagation velocity of the information. No, Mach's principle cannot be realized in this way. However, in principle we could also work within the framework of "fields". If the inertial fields are produced similarly as the gravitational fields, i.e., by the cosmic mass distribution, then we would have at each space point and at each point in time a certain value for the inertial field. The train would respond just on this value of the inertial field. However, also this conception is not maintainable because train and cosmic matter do not form a closed system [28]. If a certain event requires an emergency braking, the cosmic masses had to know that before in order to adjust the local inertial field accordingly.

We may conclude the following:

1. Mach's principle does not work within Newton's mechanics.
2. The examples above demonstrate that within Theory of Relativity Mach's principle cannot be realized.

In Appendix F we found that the effect of inertia disappears when we enter the conventional quantum theory. This seems however to be artificial because Schrödinger's equation is based on Newton's classical description. Within this classical description, the inertia-producing interaction of a body comes into play through the interaction of the body with space, or it is defined by the interaction of the body with all the other masses in the universe if Mach's principle is considered. It is relatively easy to recognize (see Appendix F) that this inertia-producing interaction must be switched off when we go over to a quantum-theoretical description of this particle (having the mass m_0), and this is because the effect of inertia is obviously switched off in the quantum-theoretical case. Such a scenario should however be hardly possible and seems to be no more than artificial.

All these problems indicate that the interpretation of the probability density $\psi^*(x, y, z, \tau)\psi(x, y, z, \tau)$ — in connection with a real material mass (Born's interpretation of usual quantum theory), which is embedded in space — seems to be an ill construction. This is the case for the conventional quantum theory. The problems disappear when we enter the projection theory where no real material body is embedded in space and time.

4.2 AN EXTENDED FORM OF MACH'S PRINCIPLE

Mach denied that the space is the source for physically real effects, and this is because we may not consider space as a real physical entity because the elements of space are not observable. Space (space-time) should not be considered as a real physical entity like matter (Chap. 1). On the other hand, Mach's principle does not forbid using space as a container in which matter is embedded.

How can a real something (matter etc.) be embedded in space (space-time) when space cannot be considered as a real physical entity? Such a conception should be eliminated. No doubt, space and time may not be the source of physically real effects, but we should additionally require that real matter cannot be embedded in space (space-time). This is in fact fulfilled in the projection theory. In other words, within the projection theory we work with an extended form of Mach's principle.

4.3 INERTIA WITHIN PROJECTION THEORY

Eliminating space (space-time) as a real physical entity and expressing the inertia of a body in terms of interactions with the body's environment is possibly not realizable within the framework of the container principle. This supposition is supported in a certain sense when we try to understand the effect of inertia in connection with quantum phenomena.

When we apply the basic ideas of Mach's principle within the container theory, problems come up. Some of them have been pointed out in Sec. 4.1. Just within the General Theory of Relativity one expects that Mach's principle is fulfilled. This is however not the case, as we have discussed in Appendix B. Furthermore, it is hardly imaginable that a sudden train stop is correlated to the cosmic masses producing inertia (Sec. 4.1); information from the galaxies to us would take thousands or even millions of years. The reason for all that is possibly the "container principle", which has been rejected in Chap. 1, and convincing arguments justified this step.

In fact, when we work within the "projection principle" the effect of inertia can be explained more satisfactorily. Let us repeat the basic ideas of the projection principle, and let us do that on the basis of Fig. 14.

All observers are caught in space and time. Since space and time, i.e., the elements x, y, z and time τ, cannot not appear in the reality outside, we have to construct other variables when we try to construct theoretical conceptions for the reality outside. We mark these new variables by the letters R, S, T, Q. As said, the quantities R, S, T, Q have to be constructed on the basis of the space-time elements x, y, z and time τ because we are caught in space and time. It turned out that the variables R, S, T, Q are expressed by the momenta p_x, p_y, p_z and the energy E (see in particular [1] and Appendix F).

The next step in Fig. 14 (projection principle) is to develop a world view with the help of the variables p_x, p_y, p_z, E.

The results for this world view have to be projected on space-time with x, y, z and time τ and we obtain a certain image. This image would also reflect our impressions that we have in front of us in everyday life, but also those facts which have been determined with an experimental setup based on a theoretical concept.

Note that the images (as, for example, Fig. 1) are given at time τ, but in the projection theory a system-specific time t appears (Appendix F); at time τ one of the t-configurations are selected ($t = \tau$). In the projection theory we have at time τ definite probability distributions for the variables x, y, z, t and p_x, p_y, p_z, E.

The time τ is the reference time and is measured with our clocks. The system-specific time t is defined by the system under investigation and is different for different systems. The reference time τ has nothing to do with the system under investigation. In conventional physics only τ is known, but not the system-specific time t.

In Appendix F we have combined the variables x, y, z, t and marked the corresponding space-time as $(\mathbf{r}, t)$-space; on the other hand, the variables p_x, p_y, p_z, E are expressed by $(\mathbf{p}, E)$-space. In the projection theory, the interactions between the entities are defined in $(\mathbf{p}, E)$-space and are described by $\mathbf{p}, E$-fluctuations. These $\mathbf{p}, E$-fluctuations appear as correlations between the various geometrical positions in $(\mathbf{p}, E)$-space.

What are the space-time properties of a body, which does not interact with its environment? Is the law of inertia that is expressed by Eq. (9) valid? In other words, does the body move through $(\mathbf{r}, t)$-space with constant velocity (where the velocity of zero is included)? It definitely does not. But how does the body move through space and time? The answer is given in Appendix F: The body moves arbitrarily! It is at time $\tau_1 = \tau_P$ at a certain point in space, characterized by x_1, y_1, z_1, and it is at the same time τ_1 at a certain time $t = t_1$, but it is completely uncertain where t_1 is positioned on the t-scale, in the past, present or in the future. The symbol τ_P indicates that we make our observation in the present.

Within conventional classical physics the bodies move on trajectories. Let us consider a classical particle, which moves within a two-dimensional space in the x, y-direction and a velocity of $\mathbf{v} = (v_x, v_y)$; it moves from one space-position to another defining a space-distance of Δ_{xy}. The situation is depicted in Fig. 30. If this process takes place within an infinitesimal time interval $\Delta\tau = \varepsilon_\tau$, the space-interval Δ_{xy} must be infinitesimal, otherwise we will not be able to define a reasonable classical velocity $\mathbf{v} = (v_x, v_y)$ and the body would not move on a trajectory.

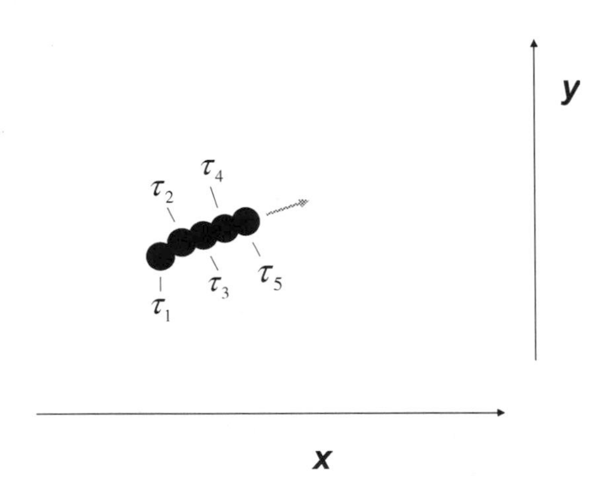

Fig. 30 Classical particles move on trajectories with well defined velocities. Let us assume that a body moves in x, y-direction from one position to another, defining a space-distance of Δ_{xy}. If this process takes place within an infinitesimal time interval $\Delta\tau = \varepsilon_\tau$, also the space-interval Δ_{xy} must be infinitesimal. This property is expressed by equation (10). The space (space-time) is here the source of inertia. In the figure the position of a body is given for times $\tau_1, \tau_2, \tau_3, \tau_4$ and τ_5 with $\tau_1 < \tau_2, < \tau_3, < \tau_4, < \tau_5$. The body moves in conventional classical physics continuously through space. This exactly corresponds to our observations in everyday life.

The Special Theory of Relativity does not allow the velocity $\mathbf{v} = (v_x, v_y)$ to be equal or larger than the velocity of light c.

If the body has at time τ the position x, y and at time $\tau + \varepsilon_\tau$ at the position $x + \varepsilon_\tau$, $y + \varepsilon_y$ we have in conventional classical mechanics the following relationships:

$$\begin{aligned} \tau &: x, y \\ \tau + \varepsilon_\tau &: x + \varepsilon_x, y + \varepsilon_y \\ \mathbf{v} &= (v_x, v_y) < c, \end{aligned} \qquad (10)$$

where ε_τ and ε_x are infinitesimal quantities. In this case, the body moves relative to space, and its $\mathbf{p}$, E-state is dependent on its velocity $\mathbf{v} = (v_x, v_y)$ relative to space, i.e., the variation of $\mathbf{v} = (v_x, v_y)$ leads to changes of the body's $\mathbf{p}$, E-states (within Newton's mechanics we have $\mathbf{p} = m_0\mathbf{v}$, $E = m_0\mathbf{v}^2/2$).

In the projection theory the situation is different. Here we have an additional variable; it is the system-specific time t, which is not known in conventional physics. Again, the time τ is used in the projection theory as an external parameter and is measured, as in conventional physics, by our clocks.

When we go from the usual quantum theory to the projection theory we have the transition (see Eq. (F5))

$$\mathbf{r}, \tau \rightarrow \mathbf{r}, t, \tau.$$

The additional variable t is important for the basic understanding of that what we call time. At time τ we have probability distributions $\{\mathbf{r}\}$, $\{t\}$, $\{\mathbf{p}\}$, $\{E\}$ for the variables $\mathbf{p}$, E, r, t: In usual quantum theory we have only two probability distributions $\{\mathbf{r}\}$, $\{\mathbf{p}\}$; the energy E remains a certain quantity in the usual quantum theory.

Bodies that do not interact with external systems, behave in projection theory completely statistically in $(\mathbf{r}, t)$-space, i.e., with respect to the variables $\mathbf{r} = (x, y, z)$ and t. (Concerning the interaction and configuration, the specific conditions outlined in Sec. F.11 are relevant.) We may define velocities $v_\mathbf{r}$ relative to x, y, z, and we have velocities v_t relative to the t. It is pointed out in Appendix F that the values of such velocities are not limited (Eq. (F66)):

$$v_{\mathbf{r}} \to \infty, v_t \to \infty$$

In the projection theory, instead of Eq. (10), we have the following scheme:

$$\begin{gathered} \tau : x,\ y, t \\ \tau + \varepsilon_\tau : x + \Delta x,\ y + \Delta y, t + \Delta t, -\infty \leq \Delta x, \Delta t \leq \infty \\ v_{\mathbf{r}} = \mathbf{v}(v_x, v_y) \to \infty, v_t \to \infty \end{gathered} \tag{11}$$

Here the bodies do not move continuously through space and time as in conventional classical mechanics. The $\mathbf{p}, E$-state of a body is "not" dependent on its velocity $\mathbf{v} = (v_x, v_y)$ relative to space, i.e., the variation of $\mathbf{v} = (v_x, v_y)$ does "not" lead to changes of the body's $\mathbf{p}, E$-states. There is a big difference to conventional classical mechanics: Here the $\mathbf{p}, E$-state of a body is dependent on its velocity $\mathbf{v} = (v_x, v_y)$ relative to space, i.e., the variation of $\mathbf{v} = (v_x, v_y)$ leads to changes of the body's $\mathbf{p}, E$-states because we have $\mathbf{p} = m_0\mathbf{v}$ and $E = m_0\mathbf{v}^2/2$.

Concerning Eq. (11) we have the following situation: At time τ the body takes the values x, y and t; at time $\tau + \varepsilon_\tau$ it may be anywhere in space and time ($-\infty \leq \Delta x, \Delta t \leq \infty$), where ε_τ is again an infinitesimal quantity. There are no restrictions. The body jumps arbitrarily through space and time (see in particular Fig. 31). The situation, expressed by Eq. (11), can be defined as the "ground state" of the projection theory. Condition for the ground state: The body does not interact with external systems (for example, via a distance-dependent potential). The ground state of conventional classical physics is defined by $v = \mathbf{v} = const$ (Eq. (9)).

In Fig. 31(a) the space position of a body is given for times $\tau_1, \tau_2, \tau_3, \tau_4, \tau_5$ and τ_5 with $\tau_1 < \tau_2, < \tau_3, < \tau_4, < \tau_5$. In contrast to classical mechanics (Fig. 30), the body jumps arbitrarily through space. In particular, its velocity $v_{\mathbf{r}} = \mathbf{v} = (v_x, v_y)$ may be infinite and can be larger than the velocity of light. In Fig. 31(b) the body moves arbitrarily relative to the time axis t without having to define a certain direction; we have at various reference times τ_i the following

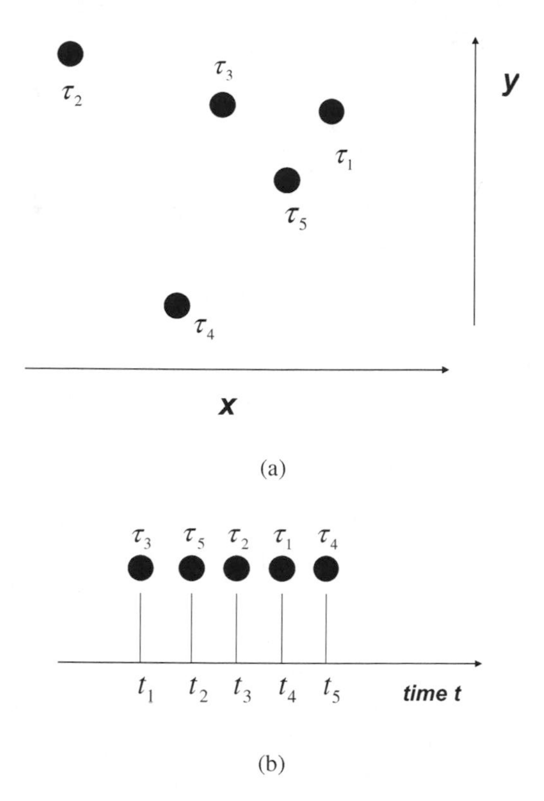

Fig. 31 Bodies that do not interact with external systems, behave completely statistically in $(\mathbf{r}, t)$-space. Here the bodies do not move continuously through space and time. At time τ the body takes the values x, y and t; at time $\tau + \varepsilon_\tau$ it may be anywhere in space and time. We may define effective velocities $\upsilon_\mathbf{r}$ relative x, y, z and also effective velocities υ_t relative to t. In figure (a) the space position of a body is given for times $\tau_1, \tau_2, \tau_3, \tau_4, \tau_5$ and τ_5 with $\tau_1 < \tau_2, < \tau_3, < \tau_4, < \tau_5$. The body moves arbitrarily through x, y-space. In figure (b) the body moves arbitrarily relative to the time axis t without to define a certain direction. In this example we run through $(\mathbf{r}, t)$-space with the following sequence $t_4 \rightarrow t_3 \rightarrow t_1 \rightarrow t_5 \rightarrow t_2$ with $t_1 < t_2 < t_3 < t_4 < t_5$ and $\tau_1 < \tau_2, < \tau_3, < \tau_4, < \tau_5$.

system-specific times $t_i, i = 1, \cdots, 5,$:

$$\begin{aligned}
\tau_1 &\rightarrow t_4 \\
\tau_2 &\rightarrow t_3 \\
\tau_3 &\rightarrow t_1 \\
\tau_4 &\rightarrow t_5 \\
\tau_5 &\rightarrow t_2
\end{aligned}$$

In this example we run through $(\mathbf{r}, t)$-space with the following sequence

$$t_4 \rightarrow t_3 \rightarrow t_1 \rightarrow t_5 \rightarrow t_2$$

with $t_1 < t_2 < t_3 < t_4 < t_5$ and $\tau_1 < \tau_2, < \tau_3, < \tau_4, < \tau_5$.

4.4 DARK MATTER IN THE PROJECTION THEORY

We may assume that the values of $\tau_1, \tau_2, \tau_3, \tau_4$ and τ_5 of the reference time are close together and are between a certain time interval A_Δ:

$$\tau_P \leq \tau_1, \tau_2, \tau_3, \tau_{4_5}, \tau_5 \leq \tau_P + A_\Delta$$

where τ_P is again the reference time in the present (see Fig. 32). Then, the times t in the interval A_Δ can be defined in the present if A_Δ is sufficiently small. For times $t < \tau_P$, the body is in the past, and for times $t > \tau_P + A_\Delta$ the body occupies the future.

In conventional physics matter really exists only in the present; within Newton's mechanics a body moves through space as a function of τ. Here we have a three-dimensional entity, i.e., we have a (x, y, z)-block. The world (system) only exists in the present, but not in the past and the future. In conventional physics the world (system) moves from the past over the present to the future. If the time τ_P defines the present, the world (system) exists for $\tau = \tau_P$ but not for $\tau < \tau_P$ and $t > \tau_P + A_\Delta$. This behaviour is directly coupled to consciousness; it is perceived at $\tau = \tau_P$ but not for $\tau < \tau_P$ and $\tau > \tau_P + A_\Delta$.

The situation in the projection theory is different from that in conventional physics. In the projection theory matter also exists for times $t < \tau_P$ and $t > \tau_P + A_\Delta$, but we actually have only those t-states of matter, which occupies the present ($\tau_P < t < \tau_P + A_\Delta$). Thus, we may classify those parts of matter that exists for $t < \tau_P$ and $t > \tau_P + A_\Delta$ as "dark matter" (see Fig. 32).

In the projection theory the description of matter-states normally takes place in space and time, although the actual interaction is a fact of $(\mathbf{p}, E)$-space. Since this description is a function of x, y, z and t, the bodies in the present are essentially influenced by "dark matter" as

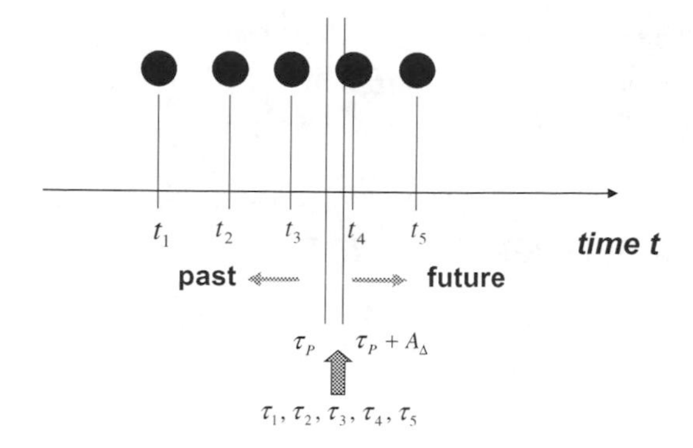

Fig. 32 The values of $\tau_1, \tau_2, \tau_3, \tau_4$ and τ_5 are between a certain time interval A_Δ, i.e., we have $\tau_P \leq \tau_1, \tau_2, \tau_3, \tau_{4_5}, \tau_5 \leq \tau_P + A_\Delta$. Then, the times t in the interval $A_\Delta(\tau_P < t < \tau_P + A_\Delta)$ can be defined as present if A_Δ is sufficiently small. For times $t < \tau_P$, the body is in the past, and for times $t > \tau_P + A_\Delta$ the body occupies the future.

well. In other words, it is not only a function of x, y, z alone. That is, the bodies in the present are influenced at time τ by bodies that occupy the past and the future at the same time τ; they do not only interact with those bodies that are positioned in the present. The "state of matter" cannot be judged on the material states in the present alone, but the whole scenario in the interval $-\infty \leq x, y, z, t \leq \infty$, consisting of past, present and future, is relevant and all parts are indivisibly connected to each other. This is a general statement and is not restricted on our example.

Let us repeat that it is useful to define "ground states". The ground state (state without external interaction) is defined in the projection theory by infinite velocities ($\upsilon_{x\,y}, \upsilon_t \to \infty$) with respect to the variables $\mathbf{r}$ and t of $(\mathbf{r}, t)$-space.

The ground state is defined in conventional physics by $\upsilon = const$ (law of inertia, Eq. (9)) and is postulated here without having to explain the mechanism; it is simply assumed that there is an interaction between space and body.

The ground state of the projection theory (expressed by υ_r, $\upsilon_t \to \infty$, see Eq. (11)) can only be reached when there are no external interactions acting on the body. In contrast to conventional classical physics the state $\upsilon = const$ must be constructed by means of the principles of the projection theory, i.e., the state $v = const$ does not belong to the basics itself. In a nutshell, inertia does not belong do the basics of the projection theory. The ground state is without inertia.

4.5 TRAVELLING THROUGH SPACE AND TIME

If we were able to control the interaction (defined in $(\mathbf{p}, E)$-space) of a spaceship with other external bodies (planets, suns, galaxies), we could reach any place in the universe at any time, that is, we would not only be able to visit any space position in the cosmos, but we could also travel in time and could visit any point in time in the past, present and future. The features, expressed by Eq. (11), allow that. The necessary condition for such space-time travels is that the crew of a spaceship is able to regulate the switching-off and switching-on of interactions in a controlled manner.

In conventional physics, the possibilities for travelling through space are formulated by Eq. (10). We know in this case that only restricted space regions can be reached within reasonable time intervals, and this is because the velocity cannot exceed the velocity of light. In connection with time-travelling certain contradictions come into play within conventional classical physics [2].

4.6 SUMMARY

1. In conventional classical physics the notion of "inertia" is of particular relevance; all real bodies are embedded in space. Even when a body does not interact with other systems, the effect of inertia is in effect and is due to the interaction of the body with space (space-time). However, space and time should not be the source of physically real effects as, for example the effect of inertia. In fact, we can never observe such a space (space-time) having the elements x, y, z and τ.

   This was the reason why Ernst Mach required the following principle: A body does not move relative to space, but relative to the centre of all the other bodies (masses) in the universe. This requirement is often discussed in literature under the notion of "Mach's principle". However, Mach's principle is not fulfilled in Newton's theory. This is also the case in the Theory of Relativity. We may conclude the following:

— Mach's principle does not work within Newton's mechanics.
— The examples given in Sec. 4.1 demonstrate that within the Theory of Relativity Mach's principle can hardly be realized.

2. When we assume that Mach's principle is valid, the inertia of a particle comes into play through the motion of the particle relative to the center of all the other masses in the universe. This inertia-producing interaction must be switched off when we go over to a quantum-theoretical description of this particle (having the mass m_0), and this is because the effect of inertia is obviously switched off in the quantum-theoretical case. Such a scenario should however be hardly possible and seems to be no more than artificial.
3. According to Mach, space (space-time) should not be considered as a real physical entity like matter. However, Mach's principle does not forbid using space as a container in which matter is embedded. How can a real something (matter etc.) be embedded in space (space-time) when space cannot be considered as a real physical entity? No doubt, space and time may not be the source of physically real effects, but we should require in addition that real matter can principally not be embedded in space (space-time). This is in fact fulfilled in the projection theory. In other words, within the projection theory we work with an extended form of Mach's principle.
4. Eliminating space (space-time) as a real physical entity and expressing the inertia of a body in terms of interactions with the body's environment is possibly not realizable within the framework of the container principle. When we work within the "projection principle" the effect of inertia can be explained more satisfactorily.
5. Bodies that do not interact with external systems, behave completely statistically in $(\mathbf{r}, t)$-space, i.e., with respect to the variables

$\mathbf{r} = (x, y, z)$ and t. We may define velocities $v_{\mathbf{r}}$ relative to x, y, z and velocities v_t relative to t. The values of such velocities are not limited and we have $v_{\mathbf{r}} \rightarrow \infty$, $v_t \rightarrow \infty$.

In the projection theory the bodies do not move continuously through space and time as in conventional classical mechanics. The $\mathbf{p}, E$-state of a body is "not" dependent on its velocity $v_{\mathbf{r}} = \mathbf{v} = (v_x, v_y)$ relative to space, i.e., the variation of $v_{\mathbf{r}} = \mathbf{v} = (v_x, v_y)$ does "not" lead to changes of the body's $\mathbf{p}, E$-states. There is a big difference to conventional classical mechanics: Here the $\mathbf{p}, E$-state of a body is dependent on its velocity $\mathbf{v} = (v_x, v_y)$ relative to space, i.e., the variation of $\mathbf{v} = (v_x, v_y)$ leads to changes of the body's $\mathbf{p}, E$-states because we have $\mathbf{p} = m_0\mathbf{v}$ and $E = m_0\mathbf{v}^2/2$.

6. In conventional physics matter really exists only in the present; within Newton's mechanics a body moves through space as a function of τ. Here we have a three-dimensional entity, i.e., we have a (x, y, z)-block. The world (system) only exists in the present, but not in the past and the future; in conventional physics the world (system) moves from the past over the present to the future. The world (system) exists always in the present, but not simultaneously in the past and in the future.

   The situation in the projection theory is different from that of conventional physics. In the projection theory matter also exists for times t in the past and for times t in the future, but we actually observe only those t-states of matter, which occupies the present. Thus, we may classify those parts of matter that exists in the past and in the future as "dark matter".

   In projection theory the description of matter-states normally takes place in space and time, although the actual interaction is a fact of $(\mathbf{p}, E)$-space. Since this description is a function of x, y, z and t, the bodies in the present are essentially influenced by "dark matter" as well. In other words, it is not only a function of x, y, z alone. That is, the bodies in the present are influenced at time τ by bodies that occupy the past and the future at the same time τ; they do not only interact with those bodies that are positioned in the present.

7. If the interaction of a spaceship (taking place in $(\mathbf{p}, E)$-space) with other external bodies (planets, suns, galaxies) could be controlled, any place in the universe at any time would be reachable, i.e., we would not only be able to visit any space position in the cosmos, but we could also travel in time and could visit any point in the past, present and future. It is known in conventional physics that only restricted space regions can be reached within reasonable time intervals, and this is because the velocity cannot exceed the velocity of light.

8. It is useful to define "ground states". The ground state (state without external potential) is defined in the projection theory by infinite velocities ($\upsilon_r, \upsilon_t \rightarrow \infty$) with respect to the variables $\mathbf{r}$ and t of $(\mathbf{r}, t)$-space. In conventional physics the ground state is defined by $\upsilon = const$. However, the ground state of the projection theory (expressed by $\upsilon_r, \upsilon_t \rightarrow \infty$), can only be reached when there are no external interactions acting on the body.

Appendix A

THE NATURE OF OUR DIRECT IMPRESSIONS

A.1 INFLUENCE OF THE BRAIN

All things, which directly appear in front of us, are positioned inside the head. We only have the impression that everything is positioned in a space outside us. Persons, cars, aeroplanes, the sun, moon and stars are images of reality in our brain; we have only the impression that all these things are located outside us.

This conclusion is supported in [29]: *We have devices in the cerebral cortex which — comparable with a television screen — produce "pictures" in our awareness from the nerve-excitations coming from the retina. It is characteristic for the sight-process that our awareness does not register the picture of a candle on the retina inside the eye, but we have the impression that we are standing opposite the candle-light which is located in the space outside, not standing on its head but upright. We see "real objects" in front of us and around us. Within this act of perception, the eye, the optic nerves and the brain work together. To see without the brain is as impossible as to see without eyes.*

This in particular means that the spontaneous impressions in front of us, which appear within assumption-less observations in everyday life (see, for example, Fig. 1), reflect observer-dependent facts.

A.2 THE COLOR-EXPERIMENT

One might believe that all the things (hard objects like houses, trees, cars, etc.) in front of us belong directly to the world outside. But this is definitely not true! All what we have in front of us is designed by the brain; it is an invention of our brain and is merely a picture of reality but is not reality or basic reality itself. This can be demonstrated by a simple reproducible experiment (see also [23] and [26]). Here are the details:

If we look in good lighting at a black ground with, for example, three circular colored areas on it (the figure is represented in [23]) approximately one minute and then look at a white surface (for example, a wallpaper), these figures will appear on this surface although they do not exist there. Furthermore, these figures do not have their original colors; instead they appear as complementary to the original colors on the black field. The figures on the white surface do not actually exist in the way observed by us anywhere, unless by chance.

The forms of the figures (three full circles) on the white and the black surfaces are the same, but only in the form, not in the color. It is relatively easy to explain this effect. In [16] we find: *The phenomenon is due to a sign of fatigue in specific color receptors on our retina. The impression "white" arises by the uniform excitation of all color receptors. If we stare at the object for an extended time, some of these color receptors will tire leaving the activity of those receptors which have not been strained to predominate and therefore allowing their color value to dominate temporarily the foreground we are experiencing.*

The color-experiment is reproducible. Under normal conditions anybody is able to experience this phenomenon so that to this extent the guarantee of objectivity is given.

What is outside, what is inside? We see the colored figure *outside* on the white surface. The explanation of the effect is however given in terms of color receptors which are inside the head. This means that the figures cannot be on a surface *outside* the observer. This also means that everything (white surface, figures, and everything else that appears in front of us) is *inside* the head. We only have the *impression* that everything is positioned in a space *outside* us.

A.3 EXPERIMENT WITH DISTORTING GLASSES

The following experiment with distorting glasses confirms that what we have found in connection with the color-experiment: All what we have in front of us is essentially influenced by the brain; it is an invention of our brain and is merely a picture of reality but is not reality or basic reality itself. The experiment itself is simple. Here the details: If one uses glasses, which strongly distort the picture in space, the following takes place: After a certain time the observer sees everything in the normal order, that is, the spatial situation is the same as before without distorted glasses and space is again right-angled (Euclidean). In other words, the distorting glasses are effectively ignored after a certain time by the observer's cognition apparatus.

Clearly, distorting glasses transform the space properties (its metric), which satisfy the axioms of Euclidean geometry, into a space with non-Euclidean geometry. Without changing the physical conditions we observe after a certain time the following strange effect: The cognition apparatus of the observer obviously transforms space with non-Euclidean geometry into usual space which satisfies the axioms of Euclidean geometry. In other words, the cognition apparatus of the observer is able to influence the space and picture, respectively.

The real physical process outside remains unchanged, only the kind of perception has been changed. We have two geometries for one and the same process outside indicating that the impression which we feel to be outside is an image within the brain of the observer.

Without doubt, the experiment with distorting glasses is basic with respect to the nature of space and the role of the observer. In particular, it demonstrates convincingly that the world in front of us is actually inside the head. This result is in accordance with that what we have concluded in connection with the color-experiment (Sec. A.2). The following experiment confirms that.

A.4 EXPERIMENT WITH INVERTING LENSES

The impression in front of us which we observe within assumptionless observations in everyday life come into existence through our brain. We need of course information from the world outside but the final impression in front of us, before our eyes, is essentially a product of mind. The color-experiment as well as the experiment with the distorting glasses confirms that convincingly. There is a further relevant experiment similar to the experiment with the distorting glasses; it is an experiment with inverting lenses and has been quoted by Thomas Kuhn (1922–1996) in [30].

An experimental subject who puts on goggles with inverting lenses initially sees the entire world upside down. At the start his perceptual apparatus functions as it had been trained to function in the absence of goggles, and the result is extreme disorientation, an acute personal crises. But after the subject has begun to learn to deal with the new world, the entire visual field flips over, usually after an inverting period in which vision is simply confused. Thereafter, objects are again seen as they had been before the goggles were put on. The assimilation of a previously anomalous visual field has reacted upon and changed the field itself. Literally as well metaphorically, the man accustomed to the inverting lenses has undergone a revolutionary transformation of vision.

Note that this visual transformation took place without the conscious action of the observer. That is, it took place within the frame of what we have called assumption-less observations in everyday life. Essentially, the spontaneous impressions that we have directly in front of us, can be influenced by the brain functions and, therefore, they are not observer-independent.

Appendix B

BASIC FEATURES OF SPACE AND TIME

Within Newton's mechanics as well as within both forms of the Theory of Relativity, space (space-time) plays the role of a container in which the real entities of the world are embedded and space (space-time) has the capability to perform physically real processes. In particular, the absolute space (space-time), an undesirable peculiarity of Newton's theory, could not be eliminated within the Theory of Relativity. Einstein wanted to overcome this problem, but his theories still contain the absoluteness of space-time.

Within the Special Theory of Relativity the situation is relatively simple: Newton's absolute space is merely extended to an absolute space-time.

Within the General Theory of Relativity we have a similar situation, but not so apparent. In this Appendix we would like to demonstrate that even within the General Theory of Relativity space and time still have problematic peculiarities analogous to those which appear in Newton's mechanics. All these facts, which belong to these undesirable and problematic features, have already been quoted in [1], [2] and [22]. Because of the fundamental importance, we want to repeat the arguments.

First, we would like to list the main facts about the nature of space and time. After that we will discuss the possibility for an empty space-time within the General Theory of Relativity, Gödel's universe and, furthermore, we will also discuss, within the General Theory of Relativity, the phenomenon of inertia in connection with a lone particle in the universe.

B.1 FACTS

Within the General Theory of Relativity there is no clear line between space-time and matter as it should be. From certain formulations in literature we may conclude that for some scientists there is no line at all: The "warp in space" and "matter" have often been considered as different aspects of the same thing. Instead we have to decipher the "code" that is anchored in the relationship between "space-time" and "matter", and this relationship is given by the following facts (see also Sec. 1.4):

Feature 1: We definitely cannot see, hear, smell, or taste space and time; that is, space and time (absolute or non-absolute) are not accessible to our senses. Also measuring instruments for the experimental determination of the space-time points x, y, z, and τ are not known and even unthinkable.

Feature 2: We can only say something about "distances in connection with masses" and "time intervals in connection with physical processes". We cannot put a piece of space and/or a piece of time on the table.

All these facts should not be fulfilled when we develop a theoretical conception of the world. This is not fulfilled within Newton's theory but also not within the General Theory of Relativity. This is demonstrated in this Appendix.

Remarks concerning feature 1

Space and time should never be the source for physically real effects as, for example, inertia. This is not fulfilled within the General Theory of Relativity: Within Einstein's theory it is still possible to talk about the rotation of the entire mass of the world relative to absolute space (Gödel's universe, see Sec. B.5 and [31]), and in this case the absolute space must be the source of inertia.

Remarks concerning feature 2

a. From feature 2 directly follows that "matter" and "space-time" are closely linked; neither should be able to exist without the other. In other words, an *empty* space-time as a physical-theoretical conception should not be existent.
b. When we define a certain distance in space by two material bodies, then there must be a relationship between the two masses and space. The material bodies can have, for example, a constant space distance which already expresses a certain kind of relationship. But what kind of relationship? This cannot be due to an interaction between the material bodies and space (space coordinates). The reason is given by feature 1. There can be no interaction between space-time and the sense organs or measuring instruments which are made of material bodies, i.e., there can be no interaction between material bodies (masses) and space-time. But what kind of relationship exists between the two material bodies that determine a certain space distance?

To sum up, space and time are not accessible by the human senses, and no measuring instrument is able to make space-time-specific "clicks" but, on the other hand, we experience space-time in connection with material bodies and processes as a concrete phenomenon. Thus, we might assume that space-time has to be identified with a "substratum". However, this idea has never been considered within the framework of serious scientific argumentations. In Sec. B.8 we will briefly discuss Einstein's thoughts on a space-time-filling substratum, different from a substratum describing space-time.

B.2 EMPTY SPACE-TIME WITHIN THE GENERAL THEORY OF RELATIVITY

Evidently, no realistic space-time theory should contain an empty space-time as a solution. Einstein thought that his field equations would fulfil this important and basic condition. However, in 1917, de Sitter gave a solution to Einstein's field equations which corresponds to an empty universe, i.e., within the framework of this solution, space-time could exist without matter, and this is in obvious contradiction to Mach's Principle. This fact was annotated by Einstein's collaborator Banesh Hoffmann as follows: *Barely had Einstein taken his pioneering step when in 1917 in neutral Holland de Sitter discovered a different solution to Einstein's cosmological equation. This was embarrassing. It showed that Einstein's equation did not lead to a unique model of the universe at all. Moreover, unlike Einstein's universe, de Sitter's was empty. It thus ran counter to Einstein's belief, an outgrowth of the idea of Mach, that matter and space-time are closely linked that neither should be able to exist without the other.* [32]

An empty space-time, isolated from real reality, is however in contradiction to the basic fact (see Sec. B.1) that we can only say something about "distances in connection with masses" and "time intervals in connection with physical processes". Thus, an empty, isolated space (space-time) reflects a metaphysical entity.

B.3 THE EFFECT OF INERTIA

According to Mach space-time can never be the source for physically real effects, i.e., inertia cannot reflect a space phenomenon. But alas this condition is also not fulfilled within the General Theory of Relativity. This has been demonstrated by de Sitter. In [33] we find the following text: *In 1917, an eminent Dutch astronomer, Willem de Sitter, pointed out to Einstein that there was a finite valued solution of his field equations that gave the inertial mass of a particle even if it was the only one in the universe. In this case, the curved space-time of general relativity would be flat, that is, the geodesic line passing through the particle would be straight. The lone particle would be guided along this geodesic line as if it was made of inertial matter. Einstein initially argued strongly against this solution. However eventually he conceded that his interpretation of inertia could not therefore be due to other matter, as required by Mach's principle, because there was no other matter around in de Sitter's example.*

In essence, the phenomenon of inertia as a real space (space-time) effect is a problematic construction within Newton's mechanics, but it also appears within the General Theory of Relativity, not only in connection with a lone body in the cosmos, but also with respect to the entire cosmos (Sec. B.5).

B.4 WHERE DOES INERTIA COME FROM?

In Newton's physics, space and time are absolute quantities, they are independent from each other, and they may even exist when space is not filled with matter. Only with these space-time features Newton was able to construct a reasonable theory of motion. In other words, it is unthinkable to eliminate the absoluteness of space and time within Newton's mechanics.

However, the concept of an absolute space (and that of absolute time) has led to enormous intellectual problems which could not be eliminated by the Theory of Relativity.

Why does the concept of absolute space-time cause serious problems? Let us briefly repeat the main points. Concerning the term "absolute", note the following (Sec. 1.3.3):

1. Absolute space was invented by Newton for the explanation of inertia. However, we do not know other phenomena for which absolute space would be responsible. So, the hypothesis of absolute space can only be proved by the phenomenon (inertia) for which it has been introduced. This is unsatisfactory and artificial. Such elements should have no existence in a physical theory, i.e., they should not be introduced in science at all.
2. The term "absolute" not only means that space is physically real but also . . . *independent in its physical properties, having a physical effect, but not itself influenced by physical conditions* [5, 34]. This must also be considered as unsatisfactory and is in particular unphysical.

Both points indicate that the concept of absolute space is actually an ill construction. Although Newton's mechanics was very successful (and it is still used in many calculations) a lot of physicists could not accept the concept of absolute space. This is demonstrated by the fact that scientists tried to solve this problem again and again up to the present day.

As we already remarked above, alas within the General Theory of Relativity the absoluteness of space-time could not be eliminated. Evidence for that is particularly reflected in a specific solution of Einstein's field equation, which is known in literature as the Gödel universe. What does the "Gödel universe" mean?

B.5 GÖDEL'S UNIVERSE

Within the General Theory of Relativity it is still possible to talk about the rotation of the entire matter of the universe relative to absolute space. Gödel's solution of Einstein's field equations leads to this result. Heckmann remarked: *This solution by Gödel describes a model of the world which is uniformly filled with matter. All points in it are equivalent, which is therefore homogeneous as in the cases mentioned up to now; it is infinitely large and rotates absolutely, but is not able to expand. At the beginning, in 1916, and still long time after that, Einstein himself believed that his theory would contain the relativity of all motions. Gödel's solution was the first solid evidence that Einstein's belief was an error.... The absoluteness of space, which Newton claimed, and Einstein thought to have eliminated, is still contained in Einstein's theory, insofar as in Einstein's theory the concept of an absolute rotation is completely legitimate. Within Einstein' theory it is still possible to talk about the rotation of the entire mass of the world relative to absolute space.* [6]

B.6 FURTHER STATEMENTS CONCERNING ABSOLUTE SPACE

The discussion with respect to the notion of "absolute space (space-time)" was and is still one of the most exciting topics in physics. Concerning absolute space, Max Born wrote: *Indeed, the concept of absolute space is almost spiritualistic in character. If we ask, What is the cause for the centrifugal forces? the answer is: "Absolute space." If, however, we ask what absolute space is and in what other way it expresses itself, no one can furnish an answer other than that absolute space is the cause of centrifugal forces but has no further properties. This consideration shows that space as the cause of physical occurrences must be eliminated from the world picture.* [35]. Furthermore, we find in this connection [35]: *Sound epistemological criticism refuses to accept such made-to-order hypothesis. They are too facile and are at odds with the aims of scientific research, which is to determine criteria for distinguishing its results from dreams of fancy. If the sheet of paper on which I have just written suddenly flies up from the table, I should be free to make the hypothesis that a ghost, say the spectre of Newton, had spirited it away. But common sense leads me instead to think of a draft coming from the open window because someone is entering by the door. Even if I do not feel the draft myself, this hypothesis is reasonable because it brings the phenomenon which is to be explained into a relationship with other observable events.*

Although Mach's principle reflects a fundamental and important feature, it could not be realized within the General Theory of Relativity up to the present day. Dehnen remarked: *In those days Einstein had in mind that the structure of space and time is given completely by the particular distribution of matter in the world in accordance with his field equations of gravitation. As a result of this, Mach's idea would be fulfilled simultaneously, after which the inertia of material bodies is determined by other masses in the world ... However, it should be emphasized that Einstein's vision that Mach's principle could be realized within the framework of the general theory of relativity failed, even by additional modifications of the original field equations ... The problem in connection with the absolute space-time within the framework of special theory of relativity — a relict from Newton's mechanics — is, in the general theory of relativity, still not solved ...* [36]

Dehnen also discusses certain reasons why Mach's principle cannot be fulfilled within the General Theory of Relativity. One reason is obviously due to the fact that Einstein's field equations are non-linear in character; here we deal with non-linear differential equations, *so that the metric field becomes — due to his self-interactions — a certain independence from matter* [36].

Within the General Theory of Relativity there is no clear line between space-time and matter, and this is clearly against the facts. In the course of time, during the development of the General Theory of Relativity, space-time took on more and more the properties of matter which is in contradiction to what we can observe in connection with space and time. We can put a piece of matter on the table but not a piece of space-time. Kanitscheider remarked: *Although Einstein was led in the construction of his theory by Mach's idea of the ontological dominance of matter over space-time, he was even temporarily convinced that his theory "had taken space and time the slightest trace of an objective reality". But the opposed tendency turned out to be the case: A tendency against Mach's principle became effective by a kind of inherent self-dynamics of the theory in the course of time of the development of the conception. Space-time became more and more an ontologically respectable entity, and it took over more and more the properties of material objects* [37].

In summary, the General Theory of Relativity has a certain problem with space-time (vacuum) and this might be the reason why there are some unsolved problems in conventional physics as, for example, in connection with the cosmological constant [22]. Mach's principle is not fulfilled within the General Theory of Relativity as it should be because this principle has to be considered as an important condition for any space-time theory, and Einstein was well aware of its importance. It is more than unsatisfactory that within the General Theory of Relativity an empty space-time can exist and an absolute space-time is even definable within this theory. In the course of the development of the General Theory of Relativity, space-time took on more and more the properties of material objects; this point reflects a critical situation and can basically not be accepted.

B.7 SERIOUS MODIFICATIONS

In order to fulfil Mach's principle, serious modifications are obviously necessary. This can possibly lead to the situation where we have to abandon the actual frame of the General Theory of Relativity. For example, in the case where the gravitational constant G varies with time an additional constant of nature has to be used. This point is discussed in [38] as follows:

The idea that the strength of gravity varies is the basis of a fully relativistc theory developed by Brans and Dicke. The theory involves space-time and geodesics, and generates uniform model universes with the Robertson–Walker separation formula (6.1.3). However, the dynamical equations connecting matter with the curvature of space-time are no longer the "simplest possible", as in Einstein's theory, but involve an extra constant of nature, ω, whose effect is to make G vary with time. The introduction of ω was not an ad hoc complication of general relativity; it was hoped that the Brans–Dicke theory would satisfactorily incorporate Mach's principle into relativity (the extent to which Einstein's theory does this is still uncertain).

B.8 EINSTEIN'S SUBSTRATUM-IDEA

Einstein's motivation for the construction of the General Theory of Relativity was Mach's principle. He was always firmly convinced that space (space-time) can never be the source for physically real effects (inertia). In the case of only one particle in the universe, Einstein's field equations nevertheless allow inertial motion and the reason for that can only be the space. That is, in this case the inertia is not created by other bodies but exclusively through space (space-time). It is therefore not surprising that Einstein in 1920 discussed the possibility for the introduction of a space-filling substratum, an ether with specific properties (not in conflict with the Special Theory of Relativity, i.e., different from the electromagnetic ether he eliminated in his 1905-paper). In this section we would like to discuss some points around this topic; for this purpose let us repeat the arguments outlined in [22].

Why did Einstein propose this new sort of either? Possibly to eliminate space (space-time) as an active cause for physically real effects (inertia, etc.). But Einstein's substratum-idea never found an echo in the scientific community. Einstein never mentioned this idea after 1920. However, Einstein's introduction of such a substratum (different from usual matter) possibly reflects his uncertainty about the phenomena of space and time, even after he had developed the Theory of Relativity.

Fact is however that space-time is often considered as a new "object" that can be seen as an independent object from the outside. That is legitimate and almost imperative within the General Theory of Relativity. We have already discussed above that Einstein's field equations allow us to construct an empty space-time without matter (de Sitter universe) but this is in contradiction to all what we know about the characterization of space and time (see in particular Sec. B.1, feature 1 and feature 2).

We have to find out why. We have to decipher this code. In our opinion, any theoretical picture of the cosmos must precisely fulfil the facts reflected by feature 1 and feature 2 (Sec. B.1) but concerning this point there is a certain ignorance and carelessness, respectively. Falk and Ruppel remarked:

...*But nevertheless we are accustomed to consider space as something special, namely as a substratum, in which all the things, as objects and fields, are embedded. Space is the house which takes up physical objects, so to speak; it forms the stage where the processes take place. It seems clear to everybody what one means when he speaks about empty space, i.e., space which is free of matter and fields. But is that really the case? It is astonishing again and again how easily we adopt conceptions and naturally and inevitably we hold them. Our conception of empty space certainly belongs to this category* [39].

In the General Theory of Relativity space-time and matter are tightly connected to each other and we have the following situation: The mass is responsible for the geometry of space-time and, on the other hand, the motion of mass is determined just by this geometry. "Mass curves space-time, and space-time tells the mass how to move". In other words, there is a certain kind of symmetry between space-time and mass. Due to these features it is difficult and impossible, respectively, to consider space-time within the General Theory of Relativity as a "nothing" without any property but space-time. From this point of view, it is natural to consider space-time as a certain kind of substratum that is different from the usual matter. We already mentioned that Einstein recognized this problem and that he in 1920 discussed the possibility for the introduction of such a space-filling substratum (we already mentioned that above). Concerning this point Cole wrote:

Einstein may have buried the ether, but ironically, he also came to praise it. In fact, in his later writings, he suggests that his four-dimensional space-time comprises a new sort of ether, refurbished and rehabilitated. According to the general theory of relativity, he wrote, space without ether is unthinkable; for in such a space there not only would be no propagation of light, but also . . . no basis for space-time intervals in the physical sense. But this ether may not be thought of as endowed with the qualities of ponderable media [8].

If space-time cannot exist without such a space-time-filling substratum, we may also say that space-time has to be identified with such a substratum. Otherwise we have merely introduced a hidden substratum, different from matter, without the answer to the question what space-time really is.

However, the identification of space-time with such a substratum seems to be problematic, not to say impossible. The reason is simple: The elements of space and time are principally not observable and, therefore, the same must be true for the substratum: Its elements would also not be accessible to empirical tests, i.e., feature 1, formulated in Sec. B.1, has to be fulfilled. However, science is essentially based on observations; without them we leave the realm of natural science.

Any scientific approach may only contain "realistic" statements about space and time and, as we mentioned several times, these are given by the following facts: We can only say something about distances in connection with material objects, and times in connection with physically real processes. Only these facts have to be considered within scientific theories where space and time are involved. If we identify space and time with a "substratum" exactly the same arguments hold. Alas this substratum has to be considered as a non-observable quantity. But a substratum with such a feature is a non-scientific term and this is not acceptable at all because the definition and consideration of non-scientific, metaphysical quantities within a scientific theory cannot be accepted and is therefore unsatisfactory. For such a substratum the same arguments that we used in connection with "absolute space" are valid; see the discussion in the preceding sections of this Appendix, in particular, the critique by Max Born.

Fact is that we never observe the curvature of space directly but only what it does to matter. For example, Einstein's concept of curved space-time was first proved when light skimmed the sun during a solar eclipse. But this bending effect can also be described by Newton's theory, though it is not essential that Newton's theory can only give a qualitative description for this effect.

Matter curves space-time within the General Theory of Relativity. How does it work? What is the mechanism for that? The General Theory of Relativity does not answer this question. It is perhaps a wrong question. Within the General Theory of Relativity there is merely an "assignment" between matter and a curved space-time, and we have to ask what assignment means in this connection. Assignment simply means that there is a curvature of space-time without any mechanism (process), i.e., matter does not interact with the space-time-elements x, y, z and τ in order to produce a curvature (feature 1, Sec. B.1). We have outlined that the General Theory of Relativity suggests the existence of a substratum, that the curvature of space-time should be due to an interaction process of usual matter with this substratum. However, this would be a new kind of interaction not defined in the General Theory of Relativity.

B.9 NEWTON, LEIBNITZ, KANT

The absolute space-time of the General Theory of Relativity is a non-observable entity. Clearly, Newton's space and time are not directly visible but, nevertheless, Newton as well as Einstein assumed that space (space-time) is really out there — essential ingredients of the real world, as real as matter. However, they could not further specify this point.

Already Gottfried Wilhelm Leibnitz (1646–1716), a German philosopher and mathematician, declined Newton's view radically and argued that Newton's space and time cannot be more than an illusion and would have nothing to do with the essential ingredients of the real world. Leibnitz clearly recognized that space and time were only a way of thinking about the world, a way of perceiving the relationships between the real things in the world. Here is a clear line, a clear qualitative difference between the real world and space-time, and this is obviously more realistic than Newton's and Einstein's views because space and time are in fact principally not observable, as we have outlined above; Newton's and Einstein's views do not clearly distinguish between reality and space-time as it should be due to the basic properties of space and time.

Let us already mention here that Immanuel Kant thought exactly in the same direction as Leibnitz. However, both, Leibnitz and Kant were not be able to give mathematical formulations for their basic thoughts, that is, for the relationship between space-time and the real world. Newton's mechanics is a very well working mathematical theory and could describe and predict a lot of experimental material; this fact pushed Kant's and Leibnitz' principal ideas away into the background but up to now there has been no serious solution for this basic and essential space-time problem.

Appendix C

COSMOLOGICAL CONSTANT

C.1 ON THE FOUNDATIONS OF CONVENTIONAL PHYSICS

In the theoretical description of the universe the General Theory of Relativity and the conventional quantum theory are relevant. Local systems as, for example, the solar system, can be treated very reliably on the basis of the General Theory of Relativity. In the case of global cosmic structures the situation is not so clear and not so easy to describe. The accelerated expansion of the whole cosmos alone has initiated an extensive and critical discussion about the structure of the Einstein's field equations, and this discussion is obviously just at the beginning. The problems are by no means easy to solve; here the cosmological constant Λ is in the centre. In this connection it is important to mention that we get un-acceptable cosmic features when we estimate Λ by means of quantum field theory; serious contradictions appear, which clearly show that we are still far away from a satisfactory picture for the whole cosmos.

The reason for this unsatisfactory situation is mainly due to the fact that the General Theory of Relativity and the conventional quantum theory, bases for cosmological descriptions, are mutually incompatible. Both theories lead to sets of laws that work fantastically. However, if we put them together we inevitably obtain irreconcilable differences. In a nutshell, both theories seem to work perfectly but, on the other hand, both theories are mutually exclusive.

This is clearly reflected in the quantum-theoretical treatment of Λ which leads to a disaster when we estimate the vacuum energy by means of quantum field theory. The problem associated with the vacuum energy is specifically detailed, and this disaster can obviously not be eliminated by the invention of certain mechanisms within the frame of one of the two theories. In other words, the discrepancy in the vacuum energy and the cosmological constant obviously reflects a basic fact which has probably its source in the very foundations of the two theories.

Quantum theory and the Theory of Relativity have been developed independently of each other, and both theories differ considerably from each other in their conceptions and contents. But a realistic cosmology needs both the description of relativistic effects as well as the description of quantum phenomena. Because both theories, the General Theory of Relativity and the conventional quantum theory, differ considerably from each other, it is problematic to use the present forms of quantum theory and the Theory of Relativity as final descriptions. As already mentioned, this is particularly recognizable in connection with the concept of vacuum leading to a quantum theoretical value for the cosmological constant which is not compatible at all with the basic structure of the General Theory of Relativity. What might be the reason for that? Both theories work within the framework of the "container principle" and, as we have pointed out in this monograph, this principle reflects a physical conception that has to be considered as more than questionable. The cosmological constant describes a vacuum state, that is, the value of Λ is strongly dependent on the space-time structure used in its estimation. However, the container principle cannot be considered as a realistic space-time concept. It is therefore not surprising that the value of Λ leads to unacceptable cosmic features, and this is because the calculated value is based on theories, which work within the container principle.

C.2 QUANTUM FIELD THEORETICAL VALUE FOR Λ

No doubt, in modern cosmology the cosmological constant Λ is of particular importance. How large is it? What energy effects are involved in its determination? Does it exist at all? Let us first discuss a basic problem in connection with Λ that is often considered as the deepest mystery in physics.

It has often been argued that we cannot conclude with absolute certainty that the cosmological observations Λ is zero: $\Lambda = 0$. An estimation leads to $|\Lambda| < 3 \times 10^{-52}\, m^{-2}$. The cosmological constant can be interpreted as constant pressure which is repulsive for $\Lambda > 0$ and attractive for $\Lambda < 0$. Λ is connected to a vacuum energy density of

$$\rho(\Lambda) = \frac{\Lambda\, c^2}{8\pi\, G}, \tag{C1}$$

where G is the gravitational constant and c the velocity of light. Within the framework of quantum field theory we obtain in fact a non-vanishing energy density which is dependent on the model but should take at least the value of

$$\rho = \frac{m_P}{l_P^3}, \tag{C2}$$

where m_P is the Planck mass ($m_P = 2, 2 \times 10^{-8} kg$) and l_P is the Planck length ($l_P = 1.6 \times 10^{-35} m$). From this we obtain for Λ the value of $\Lambda = 9.8 \times 10^{70} m^2$. Thus, on the basis of this specific estimation we may conclude that the theoretical prediction deviates from the observed value by a factor of 10^{122}; this is a lot and hardly imaginable. (In literature we often find the factor 10^{120} instead of 10^{122} or values around 10^{120} depending on the estimation.)

The factor 10^{122} is in fact a mind-boggling number. Even the quotient

$$\frac{\text{mass of all atoms in the universe}}{\text{mass of one atom}}$$

does not approach 10^{122}. In fact, if the vacuum would really contain all the energy that quantum field theory predicts we would have cosmic features that have nothing to do with that what we really observe. This discrepancy reflects one of the deepest mysteries in physics.

In summary, if we combine the value of the vacuum energy resulting from usual quantum field theory with the General Theory of Relativity the cosmological constant is obviously much too large and leads to cosmic features that we do not observe. However, this argumentation is based on the assumption that the General Theory of Relativity is correct. Just the opposite could be the case: The value of the vacuum energy resulting from the usual quantum field theory is correct but not the space-time structures of the General Theory of Relativity.

C.3 WHAT IS WRONG?

The situation in connection with the cosmological constant Λ demonstrates that there is something fundamentally wrong. But what? This question is difficult to answer. The factor of 10^{122} in connection with Λ is probably due to the fact that the conceptions of the General Theory of Relativity and the conventional quantum theory not only differ considerably from each other but are in particular not compatible.

The formulas of the General Theory of Relativity and those of the conventional quantum theory work fantastically in most cases but there are still open questions in connection with some basic points. For example, the problem of "time" is not satisfactorily solved within the usual quantum theory. Here time still behaves classically within all forms of the usual quantum theory. On the other hand, Mach's Principle is not fulfilled within the General Theory of Relativity and this is weighty. According to Mach space should not be the source of physically real effects such as the phenomenon of inertia. Both points, the "nature of time" and "Mach's principle" are of fundamental importance for the description and the basic understanding of the universe, and the cosmological constant obviously plays a central role here.

We already mentioned that the General Theory of Relativity and the conventional quantum theory cannot be both right. What theory is correct and what not? Alas this question cannot be answered in a simple way. It is probably realistic to assume that both theories require modifications, although they are both able to describe certain experimental material impressively. It would be desirable when both theories (General Theory of Relativity and conventional quantum theory) could be based on the same basic principles providing a unification of gravitational and quantum effects. Where have we to search?

In both theories, the container principle is applied. Is the container principle an adequate description? Is the naive standpoint really true that the real world is "embedded" in space-time? We will see that this standpoint is questionable.

C.4 BASIC ELEMENTS FOR A VACUUM DESCRIPTION

Our knowledge about the vacuum (space) properties is presently based on the usual (conventional) quantum theory and the Special Theory of Relativity. For the estimation of the vacuum energy, conventional quantum theory delivers the uncertainty relations, in particular, the so-called uncertainty relation for the energy E and time τ:

$$\Delta E \Delta \tau \geq \frac{\hbar}{2} \tag{C3}$$

From the Special Theory of Relativity the formula $E = mc^2$, the well-known relation between energy E and mass m, is used. Both relations lead to the postulate of antimatter: To every particle there exists an antiparticle (it is, for example, a positron for an electron). Dirac actually considered the antiparticle as a very palpable hole in nothing. However, *Dirac did not have all the details right; now that physicists understand antimatter better, it is no longer viewed as holes in nothing. A positron is a real particle in its own right. Still, as Wheeler points out, the modern antiparticle theory and Dirac's hole theory differ only in their imagery, not their mathematics* [8]. Whatever the case, the particle-antiparticle concept is relevant in the estimation of the vacuum energy.

Due to Eq. (C3) the energy can fluctuate, and a particle-antiparticle pair may appear from the vacuum for a short time interval dictated by Eq. (C3). The ground state energy of the vacuum can never be zero because Heisenberg's uncertainty relation (C3) is assumed to be valid for all physical situations. Just these energy fluctuations, due to Eq. (C3), are responsible for the large vacuum energy density and the large value for the cosmological constant Λ. What can we say about the validity of Eq. (C3)? What are the limitations of this equation?

C.5 ON THE SYMMETRY BETWEEN SPACE AND TIME

Here we are interested in the symmetry between space and time. There is symmetry in the Special Theory of Relativity but not in the conventional quantum theory. Let us point out more details. Since conventional quantum theory and Special Theory of Relativity have been developed independently from each other, the following question is of relevance: Can quantum phenomena be treated fully relativistically in accordance with the basic laws of the Special Theory of Relativity? Although the relativistic wave equations (e.g. Dirac's equation for the electron) are invariant under Lorentz transformation, the space coordinates x, y, z and time τ are in its physical content definitely not symmetrical to each other and that is in contrast to the fundamental results of the Special Theory of Relativity. This is essentially due to two facts:

(i) Whereas the coordinates are "statistical" quantities, time does not behave statistically. We already mentioned above that time remains unchanged when we go from classical mechanics to the conventional quantum theory. This is clearly reflected in the fact that the coordinates can be "operators", time τ is always a simple "parameter". No doubt, this reflects a serious problem.
(ii) The determination of the eigenfunctions and eigenvalues is restricted on space; time is not involved in this process.

Louis de Broglie (1892–1987) expressed this fact as follows: *The present quantum theory in all its versions takes time as the evolution parameter and therefore destroys the symmetry between space and time* [40].

Within the framework of the Special Theory of Relativity the world has to be considered as four-dimensional, and this is because time loses its independence which it still had within Newton's mechanics. The fourth equation of the Lorentz transformations

$$\tau' = \frac{\tau - \upsilon x/c^2}{\sqrt{1 - \upsilon^2/c^2}} \tag{C4}$$

shows that the time interval between two events in a moving frame of reference S' (which moves relatively to a rest system, say S, with the constant velocity υ) does not vanish in general, even when the time interval in S becomes zero; the consequence of a pure "distance in space" in S is an "interval of time" in S', and this result can directly be read from Eq. (C4). It is well known that Eq. (C4) is realistic, and this has been experimentally demonstrated by the slowing down of clocks.

Clearly, in contrast to the usual quantum theory, within the Special Theory of Relativity space and time are tightly interrelated and symmetrical to each other and cannot be treated as independent quantities as is done in the conventional quantum theory, even in the relativistic case.

That is in close connection with the fact that there is no uncertainty relation for the energy and time which would agree in its physical content with the well-known uncertainty relation for the coordinates and momenta:

$$\delta p_x \delta x \geq \frac{\hbar}{2}, \quad \delta p_y \delta y \geq \frac{\hbar}{2}, \quad \delta p_z \delta z \geq \frac{\hbar}{2} \tag{C5}$$

An analogous relation for the time τ and a quantity which has the dimensions of energy is required from the point of view of the Special Theory of Relativity. The significance of the well-known relation (Eq. (C3))

$$\Delta E \Delta \tau \geq \frac{\hbar}{2}$$

is entirely different from that of Eq. (C5). This difference is symbolically expressed by the use of Δ instead of δ. In the relations (C5) the quantities δp_x, ... and δx, ... are the uncertainties in the values of the momenta and the coordinates at the same instant τ. As is well-known, this uncertainty means that the coordinates and momenta can never have entirely definite values simultaneously. The energy E, on the other hand, can be measured to any degree of accuracy at any instant τ. The quantity ΔE in Eq. (C3) is the difference between two exactly measured values of the energy at two different instants and is not the uncertainty in the value of the energy at a given instant (see also the discussion in [40] and [41]).

This is a strange situation and is obviously due to the fact that within usual quantum theory time τ is still a classical quantitiy. This is an important point and might be one of the keys for the understanding of the vacuum energy problem that is connected to the cosmological constant Λ (Sec. C.2). Let us cite some critical remarks with respect to $\Delta E \Delta \tau \geq \hbar/2$ (Eq. (C3)).

C.6 MARIO BUNGE'S CRITIQUE

Since time within the conventional quantum theory is still a classical parameter, we should consider the energy-time relation $\Delta E \Delta \tau \geq \hbar/2$ (Eq. (C3)) as a quasi-classical equation. The value of such an equation is however questionable.

In Mario Bunge's opinion the energy-time relation (C3) "is a total stranger to quantum theory" [42]. In particular, we find in Ref. [42]:

This relation is made plausible by reference to some thought experiments, to radioactive decay, and to line breadths. But unlike the genuine indeterminacy relations, (C3) has never been proved from first principles. In other words, (C3) does not belong to quantum theory, but is just a piece of doubtful heuristics.

The reason for this failure to incorporate (C3) into quantum mechanics is the following. In this theory, as in every other known theory, time is a "c number" and, more particularly a parameter, not a dynamical variable. Moreover, τ doesn't "belong" (refer) to the system concerned.

Even in relativistic theories the proper time, though relative to a frame of reference, is not a property of the system on the same footing as its mass or its momentum. In other words, τ does not belong to the family of operators in the Hilbert space associated to every *microsystem. Therefore, τ is not a random variable and its scatter vanishes identically... Consequently, no matter what the scatter in the energy may be, the inequality (C3) does not hold. Also, it does not improve things to regard E, as is sometimes done, as the Hamiltonian of the system. The standard deviation of the energy vanishes as well. In conclusion, the so-called fourth indeterminacy (or uncertainty) is a total stranger to the quantum theory although it can be found in works on this theory....*

In short, the fourth scatter relation is not deducible from the principles of the quantum theory, whether relativistic or not. But then why is it sometimes used, for example in the theory of line breadths? The reason is that it is not the same formula: although it has the same typographical form, it has a different content. In particular, "Δ" is not interpreted as a standard deviation (from what?) but as the half-life of the state ψ. But even thus reinterpreted, the formula is not deducible from the postulates of the general quantum theory. Mind, this has nothing to do with the question whether or not formula (C3) is true under some suitable interpretation. Many other statements are true and yet they do not belong to quantum theory.

C.7 FINAL REMARKS

The value for the cosmological constant leads to a critical situation. If we assume that the quantum field theoretical value is correct and put it into Einstein's field equation we obtain impossible cosmic features. In [8] we find the following instructive example: *Indeed, if the vacuum contained all the energy physicists expect it to, it would be so repulsive that you wouldn't be able to see your hand in front of your nose. Even at the speed of light, the light from your hand wouldn't have time to reach your eyes before the expanding universe pulled it away.*

In fact, within the General Theory of Relativity the notion "absolute space" is still defined as within Newton's mechanics (see in particular Appendix B). This could be the reason why we obtain unacceptable cosmic features. Within the Theory of Relativity space-time can be the source for physically real effects, and this is against

Mach's principle and the fact that we do not know an answer to the question "What is the space (space-time) made of?" We cannot put a piece of space (space-time) on the table. More details are given and discussed in Sec. 1.4.

However, there is of course also the possibility that the structure of the General Theory of Relativity is not the cause for the impossible cosmic features, which appear in connection with the cosmological constant. The conventional quantum theory could also be the cause for that. In fact, the time τ is still a classical parameter in the conventional quantum theory and the energy-time relation $\Delta E \Delta \tau \geq \hbar/2$ [Eq. (C3)] has to be considered as a quasi-classical equation. This is a critical point because this equation plays an essential role in the calculation of the quantum field theoretical value for the cosmological constant.

In a nutshell, both theories, i.e., the General Theory of Relativity and the conventional quantum theory, have their own deficiencies. Therefore both theories possibly have to be modified. Both theories work within the framework of the container principle, and we have pointed out in this monograph that the container conception seems to be not very realistic. Strictly speaking, it has to be considered as a non-scientific conception.

Again, is the real world really embedded in space (space-time)? From the point of view of modern physics the vacuum (the space) has nothing to do with emptiness; just the opposite is the case: Empty space (vacuum) is a "*hyperactive player, a prolific producer of jittering fields and virtual particles* [8]. *The vacuum is the most complex substance in the universe. The biggest challenge for theorists of all may simply be emptying the vacuum of all the trappings it's acquired over the past fifty years.* "*They have filled the vacuum with so much garbage, there isn't room for the cosmological constant*," said Leon Lederman: "*Einstein freed us from the ether. Now we need to get rid of (today's version of ether) again. We need to sweep the vacuum clean* [8]." No doubt, one possible solution for this problem is to work within the "projection principle" (Appendix F). In fact, here no physically real objects are embedded in space (space-time).

Furthermore, Eq. (C3) is not defined in the projection theory, i.e., the quasi-classical energy-time relation $\Delta E \Delta \tau \geq \hbar/2$ does not belong to the body in the projection theory. Instead we obtain here a genuine uncertainty relation for the energy and time [1]:

$$\delta E \delta t \geq \frac{\hbar}{2} \tag{C6}$$

where t is a real quantum time; τ in Eq. (C3) is a strict classical quantity. In contrast to Eq. (C3), relation (C6) agrees completely in its physical content with the uncertainty relation for the momentum and the coordinate. The required symmetry between space and time that is required by the Special Theory of Relativity is definitely fulfilled within the projection theory, but not in the conventional quantum theory (in its relativistic and its non-relativistic version).

Appendix D

KANT'S PHILOSOPHY

The philosopher Immanuel Kant investigated the relation between the true and the perceived reality. Kant's ideas are close to what we have developed above in Chap. 1. However, there are nevertheless big differences concerning statements about the reality outside. Let us briefly discuss Kant's ideas.

Kant argued that we cannot make statements about the true reality outside. According to him all things we observe are located within space-time and these elements, space and time, are located inside the observer.

In Kant's opinion, a human observer can say nothing about the structure of the outside world. In particular, in his opinion a human observer is not able to give answers to relevant questions: Is there a one-to-one correspondence between the structures outside and those in the brain of the observer? Does the information in the pictures, located in the brain, reflect the complete information about the reality outside? This is because Kant knew nothing about the principles of evolution, which is an essential point in answering such questions.

D.1 KANT: SPACE AND TIME ARE ELEMENTS OF THE BRAIN

According to Immanuel Kant space and time are not empirical concepts, which are determined by abstraction from experience. Experiences become possible at all only through the concepts of space and time. According to Kant space and time are not objects, but have to be considered as preconditions for the possibility of all experience. Although in Kant's opinion space and time are not empirical concepts, they nevertheless have empirical reality. This is because all things, which we observe, are located in space and time. The structure of space and time is therefore reflected in the empirical objects. Kant denied the existence of a space and a time independent of brain functions (observations in everyday life, thinking).

According to Kant space and time are located inside the observer. Whether space and time are also elements of actual (fundamental) reality outside remains principally an open question within Kant‘s point of view.

D.2 THE PHYSICAL LAWS ARE MERELY PICTURES IN THE HEAD

Kant's perspective is without any doubt important, not only in connection with philosophical questions. But what are the consequences for physics? If we take Kant's view seriously, then the physical laws as, for example, Newton's gravitational law, are merely pictures in the head of the observer and there is principally no way to express these laws for the reality outside; nothing can be said about the processes in the outside world. This is in contrast to "projection theory" outlined above (Chap. 1) where we can construct fictitious realities. If the gravitational law — and all the other physical laws — is merely a picture in the head we get a problem because there can be no gravitational forces in the head of the observer. Clearly, Kant's thoughts can lead to considerable problems when we apply them to physics. Barrow remarked: *We can see that Kant's perspective is worrying for the scientific view of the world* [4]. However, Kant's perspective has not been taken so seriously in physics and there are other positions. The situation is well analyzed by Barrow: *There are two poles about the relationship between true reality and perceived reality. At one extreme, we find 'realists', who regard the filtering of information about the world by mental categories to be a harmless complication that has no significant effect upon the character of the true reality 'out there'. Even when it makes a big difference, we can often understand enough about the cognitive processes involved to recognize when they are being biased, and make some appropriate correction. At the other extreme, we find 'anti-realists', who would deny us any knowledge of that elusive true reality at all. In between these two extremes, you will find a spectrum of compromise positions extensive enough to fill any philosopher's library: each apportions a different weight to the distortion of true reality by our senses* [4].

D.3 REALISTS AND ANTI-REALISTS

In other words, within Kant's philosophy there are no criteria to decide about the true nature of absolute reality; the realists cannot disprove the anti-realists and vice versa. The realists more or less assume that there is a one-to-one-correspondence between true reality and perceived reality (picture); the anti-realists maintain that we can say nothing about true reality. However, when we consider the basic facts of biological evolution both viewpoints seem to be not realistic. Evolution teaches us that in connection with humans and animals it is not the cognition which plays the important role in nature but the differentiation between "favorable towards survival" and "hostile towards survival", at least at the early phase of evolution ([1, 2], Sec. 3.3). Each picture of reality (perceived reality) is tailor made to this characteristic. Since the conditions for survival are different for different biological systems, the perceived realities are different for different biological systems. Wolfgang Schleidt's experiments ([1, 2], Sec. 3.3) with a turkey showed that very impressively.

The picture of reality designed unconsciously by the individual has to be correct but it may only contain, for economic reasons, information which is absolutely necessary for survival; everything else is unnecessary. The picture of reality does not have to be complete and true (in the sense of a precise reproduction) but restricted and reliable. Furthermore, we learned from Schleidt's experiments that the conception of the world of man and that of turkey are on the one hand different from each other, they are on the other hand correct in each case. This means that neither of these two conceptions of the

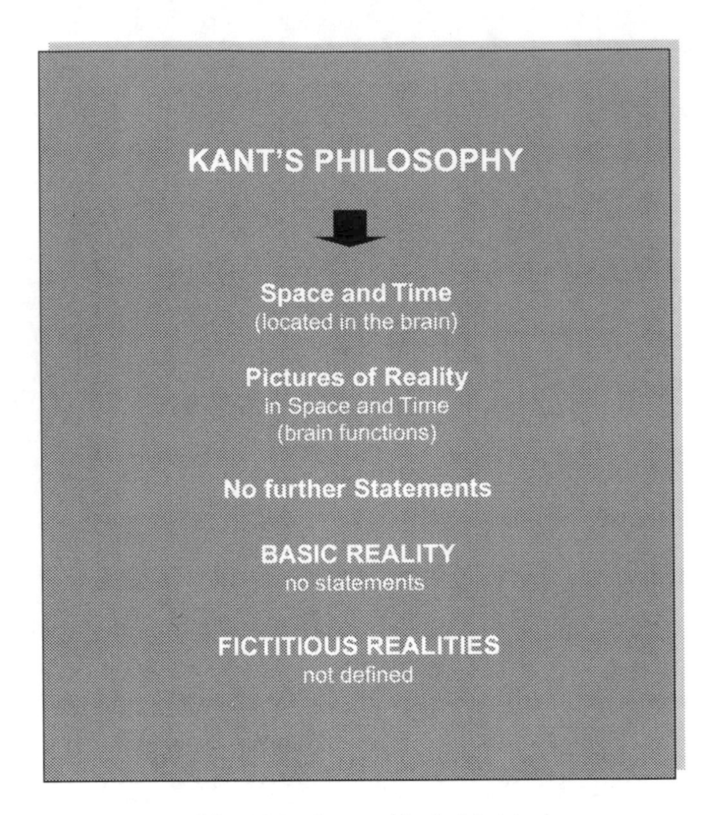

Scheme D1 Features of Kant's philosophy.

world can be true in the sense that they are a faithful reproductions of nature. There can be no one-to-one correspondence between the structures in the picture and those in true reality. Objective reality must be different from the images which biological systems construct from it.

The statement that there can be no one-to-one correspondence is, on the one hand, against the realists and, on the other hand, it is simultaneously against the position of anti-realists because it is a statement about the true reality. The main features of Kant's philosophy are summarized in Scheme D1 and those of the projection theory in Scheme D2.

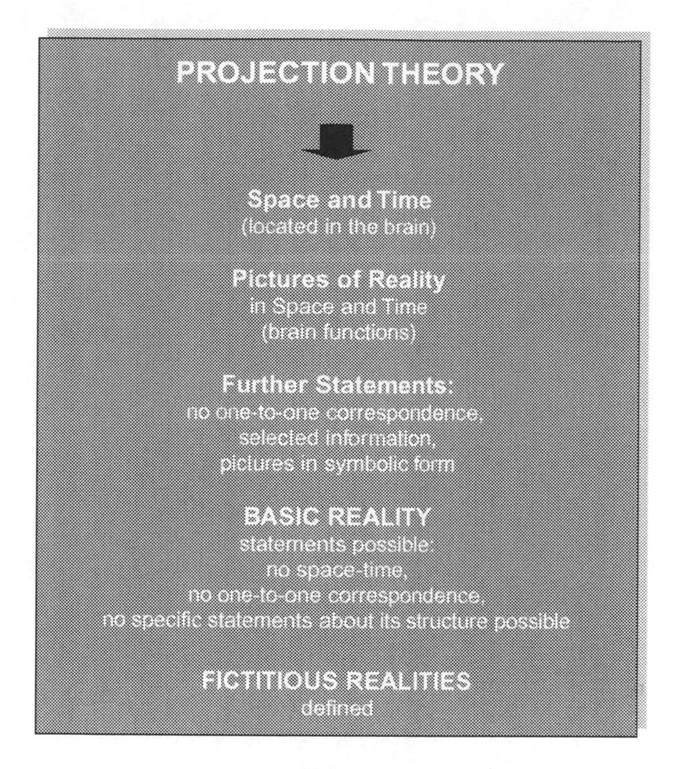

Scheme D2 Features of projection theory.

D.4 REMARKS ON NEWTON'S VIEW

Newton discovered his mechanics approximately two hundred years before Kant developed his perspective. Newton did not distinguish between the "true reality" and the "perceived reality (picture of reality)" and he based his theory directly on what was directly in front of him: Space and objects, which were more or less separated in space. The big success of Newton's theory pushed Kant's perspective into the background, and all the further developments in physics were based on Newton's view without to consider Kant's basic thoughts, which has merely remained the subject of philosophy up to the present day.

Appendix E

THEORY OF SCIENCE: BASIC STATEMENTS

Within the projection theory we can say nothing about the structure of the basic reality. We are principally not able to learn something about the absolute truth. The basic reality exists but we have no access to it.

When we work within the "theory of science", we come to the same result: Absolute truth, i.e., basic reality, must remain hidden. This conclusion is based on general arguments. What kind of arguments are used here? In this Appendix we would like to quote the main results.

We first investigate the situation within the so-called "asymptotic convergentism". This view suggests that it should be possible to reach the absolute truth after a certain time. After that we will discuss another view: Progress of science takes place by the fact that there is a fundamental change in perspective, i.e., the final, absolute truth can never be achieved.

E.1 HOW DOES SCIENCE PROGRESS?

We would like to investigate what statements about the structure of reality can be made when the peculiarities for the progress of science are considered. What laws determine this progress?

Eliminating the number of unanswered questions

Let us suppose that there exists a defined set R_S of problems which, in the course of time, will be solved successively. If we denote the number of answered questions at time τ by R and the number of answered questions at time τ' by R', we get

$$R < R' \tag{E1}$$

with $\tau < \tau'$. Accordingly, the number of questions we are able to formulate decreases:

$$(R_S - R) > (R_S - R') \tag{E2}$$

Within the frame of such a principle all questions will be answered in the course of time, i.e., a maximum of scientific knowledge is achieved. However, this statement can only be done if there is a constant number R_S of problems in the world. If that should be the case we could in principle reach the final truth R_{final} at time τ_{final} with

$$R_{final} = R_S \tag{E3}$$

Then, R_{final} reflects the complete information of the world; the complete theory is known at time τ_{final}.

Is such a transition from (E2) to (E3) possible at all? Is, in other words, the time τ_{final} reachable? We will discuss in the forthcoming sections of this Appendix that such a concept cannot be maintained. Nevertheless, Eq. (E3) reflects just that what most people, in particular scientists, are convinced of. Even in modern physics we find this tendency. For example, Steven Hawking suggested in his famous book *A Brief History of Time* that we should thoroughly be able to discover such a complete theory. He wrote: . . . *if we do discover a complete theory, it should in time be understandable in broad principle by everyone, not just a few scientists. Then, we shall all, philosophers, scientists, and just ordinary people, be able to take part in the discussion of the question why it is that we and the universe exist. If we find the answer to that, it would be the ultimate triumph of human reason — for then we would know the mind of god.*

Can a human being really be considered as godlike? I hardly think so. To assume that is rather overbearing. Instead, we have recognized in Chap. 1 that the development of a "complete world view" is principally not possible within the projection theory; within this theory the basic reality is not accessible to a human being, even specific structures within the basic reality remain hidden.

It is not only Hawking who believes that the present theoretical structures of the physical laws represent the final (absolute) truth. Let us repeat what we said in [2].

Scientists of all epochs thought that the scientific laws "now" approached their final state of completion, i.e., scientists of all epochs were convinced to have the absolute truth in their hands. However, every time it turned out that this belief was a fallacy. It is a certain kind of self-indulgence.

For instance, Lord Kelvin (1824–1907) thought that the foundations of physics as laid down towards the end of the last century were complete and that only secondary questions were still left to be answered. Berthelot (1827–1907) in 1885 felt that the world no longer concealed any secrets. Haeckel (1834–1919) concluded from his studies (also made towards the end of the 19th century) that all legitimate questions in natural science had essentially been answered. Another example of this tendency is given by Max Planck (1859–1947):

As I was beginning to study physics (in 1875) and sought advice regarding the conditions on prospects of my studies from my eminent teacher Philipp von Jolly, he depicted physics as a highly developed and virtually full-grown science, which — since the discovery of the principle of the conservation of energy had in a certain sense put the keystone in place — would soon *assume its final stable form. Perhaps in this or that corner there would still be some minor details to check out and coordinate, but the system as a whole stood relatively secure, and theoretical physics was markedly approaching that the degree of completeness which geometry, for example, had already achieved for hundreds of years. Fifty years ago (as of 1824) this was the view of a physicist who stood at the pinnacle of the times* [9].

If we jump from the last century to the present, we find a recurrence of the opinions above. For instance, Richard Feynman (1918–1988) wrote [9]:

What of the future of this adventure? What will happen ultimately? We are going along guessing the laws; how many laws are we going to have to guess? I do not know. Some of my colleagues say that this fundamental aspect of our science will go on, but I think there will certainly not be perpetual novelty, say for a thousand years. This thing cannot keep on going so that we are always going to discover more and more new laws... We are very lucky to live in an age in which we are still making discoveries. It is like the discovery of America — you only discover it once. The age in which we live is the age in which we are discovering the fundamental laws of nature, and that day will never come again. It is very exciting, it is marvellous, but this excitement will have to go. Of course in the future there will be other interests,... There will be a degeneration of ideas, just like the degeneration that great explorers feel is occurring when tourist begin moving in on a territory. In this (present) age people are (perhaps for the last time?) experiencing a delight, the tremendous delight that you get when you guess how nature will work in a situation never seen before.

It is clearly evident from all the examples cited that eminent personalities (and obviously not only these) in different epochs thought that scientific findings "now" approached their final state of completion. Evidently, there have also been epochs in which the knowledge available was judged with less self-indulgence: Nicholas Rescher in his book entitled, "The Limits of Science" [9], from which most of the texts of the examples quoted above have been taken, deals with these anti-position. However, compared with the discussions in present day literature, the opinions of the authors quoted before — especially those expressed by Hawking and Feynman — seem to dominate for the time being. However, the past showed that this belief was every time a fallacy and only a certain kind of self-indulgence remained.

The facts from the theory of science have taught us that we will never be able to make final and complete statements about the physical universe, i.e., from the point of view of the theory of science the "absolute truth" can never be obtained. This is an important statement and will be justified below.

E.2 PRINCIPLE OF PROPAGATION OF QUESTIONS

In contrast to what we have pointed out in Sec. E.1, Immanuel Kant advocated the "Principle of Propagation of Questions". This principle means that each answer to a scientific question gives rise to new questions. According to Kant this applies because each answer provides new material feeding new questions. But new questions generally change the supposition and this may lead to an extension of the problem horizon.

We no longer have a constant number R_S of problems in the world but a variable set of problems, say R'_S, is encountered. Then, an increase in the number of unanswered questions can very well be accompanied by an increase in answered questions. Since

$$R_S < R'_S \tag{E4}$$

the following relations

$$R < R' \tag{E5}$$

$$(R_S - R) < (R'_S - R') \tag{E6}$$

are possible. In other words, there are R'_S problems at time $\tau' > \tau$ and the number of answered questions is R' at τ. In this case the number of answered questions increases with time, and the number of problems R'_S increases simultaneously (see Eq. (E4)). The number of answered questions at time τ is again R and there are R_S problems at τ.

Since R'_S varies with time, it is difficult to say at which point in time the so-called "final truth" is reached; one can even not say whether such a final truth exists at all.

However, the opinions of the scientists quoted above (Lord Kelvin, Feynman, etc., Sec. E.1) can be well harmonized with Kant's "principle of question propagation" provided that not only the number of questions but also their relevance is taken into consideration. As a matter of fact, if the relevance of a problem gradually decreases in the course of time, "later" science must be less important. In that case, it can be quite correctly asserted that the basic structure of physics has been worked out although the number of questions to be answered increases with time.

Such a view of the progress of science suggests the analogy with geographic research. In this context, Rescher states among others [9]: *Scientific inquiry would thus be conceived of as analogous to terrestrial exploration, whose product — geography — yields results of continually smaller significance which fill in ever more minute gaps in our information. In such a view, later investigations yield findings of ever smaller importance, with each successive accretion making a relatively smaller contribution to what has already come to hand. The advance of science leads, step by diminished step, toward a fixed and final view of things.*

Accordingly, science progresses by approaching the truth successively: The "final answer" and the "final view" of things, respectively, is gradually approached by the way of an asymptotic approximation. According to Peirce (1839–1914) this truth is obtained in the limit $\tau \to \infty$:

$$R_{\infty} = \lim_{\tau \to \infty} R_{\tau} \tag{E7}$$

where R_{τ} is the time dependent status of knowledge which approaches asymptotically the definite truth R_{∞}. This detailed filling (accumulation) of given, fundamentally defined patterns resembles greatly the calculation of further decimal points in order to additionally refine a result already roughly estimated, such as in calculating the numerical value of π:

$$\begin{aligned} \pi_1 &= 3{,}1 \\ \pi_2 &= 3{,}14 \\ \pi_3 &= 3{,}141 \\ &\vdots \\ \pi &= \lim_{n \to \infty} \pi_n \end{aligned} \tag{E8}$$

At least until one generation ago, the opinion was firmly established that science is cumulative and the advocates of the scientific method had understood scientific progress to have this cumulative nature (see also [9]). Within this concept the "absolute truth" is set equal to "our ultimate truth". But we will see below that the asymptotic convergentism cannot be upheld in the recognition theory and scientific history.

E.3 SUBSTITUTION INSTEAD OF SUCCESSIVE REFINEMENT

All the statements cited above can be classified on the basis of the asymptotic convergentism. However, essentially two serious objections can be quoted against these statements, i.e., against the asymptotic convergentism:

1. *There is no metric to measure the "distance" between bodies of knowledge*

The asymptotic convergentism is from the beginning burdened by the great problem that it cannot give a "metric system" allowing us to define the interval concerning

$$R_\tau - R_{\tau'}$$

between the status of knowledge R_τ and the status of knowledge $R_{\tau'}$. This means that we are not in the position to decide whether or not we have approached the real truth. How can criteria be formulated which allow such an approximation, that is, to find a metric that is able to measure the "distance" of knowledge? We cannot! There is simply no neutral standpoint in theoretical terms, i.e., no neutral, elevated level (external to science) which could form the basis for a direct comparison between theoretical configurations R_τ and $R_{\tau'}$, etc. Scientific progress can only be measured in connection with the so-called "pragmatic level"; this point will be outlined in more detail below.

2. *There is a fundamental change in perspective*

The assumption of a successive approximation cannot be maintained in view of the history of science because the analysis of theories succeeding each other in time shows that the later theory is generally not only supplemented and refined, respectively, but reformulated on the basis of new first principles. A basic change in perspective took place. Normally the problem does not consist of just adding some further facts but of structuring a new frame of thinking. This situation has been described in a highly instructive manner by Thomas Kuhn (1922–1996) who compared Newton's theory to the Theory of Relativity [30]: Einstein's theory, as compared to Newton's theory, provides a basically novel frame of thinking. A change of perspective took place, involving the notational decoupling and displacement, respectively, of the network of notions.

E.4 ESTIMATES OF THE TRUTH

Generally speaking, we have to conclude that a "later" theory became necessary because the "earlier" theory had been limited in its scope. This led to a basically novel concept of the way of seeing nature. Thus, normally, not only improvements and refinements are made, but the "earlier" theory is downright replaced by the "later" one. The history of science provides a wealth of examples supporting this statement [9, 30].

E.5 ARE THERE FRAMES FOR THE "ABSOLUTE TRUTH"?

We have to assume consistently that the "later" theory will also have to be abandoned at some point in time. So, each frame of thinking, independent of the era in which it has been conceived, can never constitute a frame for the "absolute truth". In this context we cannot even provide evidence that we have come closer to the "absolute truth" because no metric system can be defined to measure intervals in recognition. A framework of thinking (theory) does not reflect the "absolute truth" but, as formulated by Rescher [9], an "estimate of the truth", which is to be understood as a tentatively postulated provisional truth.

E.6 INCOMMENSURABLE STRUCTURES OF THINKING

When changing from one to another frame of thinking, generally a conceptual decoupling takes place so that successive structures of thinking may become incommensurable. The advocates of incommensurable theories are basically not in a position to understand each other, because it is not reasonable to make comparisons between incommensurable structures of thinking. Using the following analogy, Rescher reduces the problem to the point [9]: *One can improve upon one's car by getting a better car, but one cannot improve it by getting a computer or a dishwashing machine.*

E.7 THE PRAGMATIC LEVEL

Scientific progress can be defined only if it is a possible to project certain tendencies of two incommensurable structures of thinking to one "appropriate" third level (Fig. E1). This third level, which is a sort of reference system (level *A* in Fig. E1) generally will have a "coarser" and "more global" structure than the two incommensurable structures of thinking (levels 1 and 2 in Fig. E1). Level *A* is coarser and more global because, generally, such a projection is not detailed, i.e., it is not a point by point projection and provides only an integral picture (e.g., by averaging). These integral variables of level 1 and level 2 have things in common, if they cover a joint zone on level *A*. Level *A* is "appropriate" for comparing two structures of thinking, if it offers to them a finite surface for projection. A structure of thinking is superior to another if it describes in more detail and accuracy the body of facts on level *A*.

In this way it is possible to compare (albeit to a limited extent only) two competing theories. In Fig. E1 the structures of thinking underlying level 1 is superior to that of level 2 because the surface (equivalent to the status of recognition which, starting from level 1, explores level *A*) of level 1 projected onto level *A* is larger than the surface of level 2 projected onto level *A*.

Related to the situation of man, level *A* might constitute the level of everyday life, of technological applications, and of experimental explorations. Thus, technological progress and the elucidation by experiments becomes the touchstone (albeit to a limited extent) of

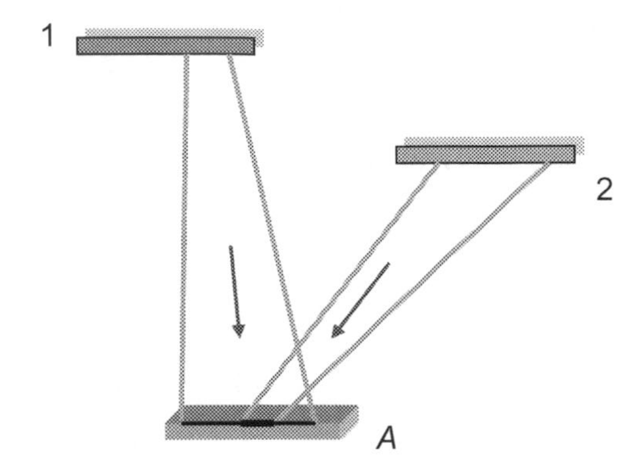

Fig. E1 Scientific progress can be defined only if it is a possible to project certain tendencies of two incommensurable structures of thinking to one "appropriate" third level (level A). The two incommensurable theories are positioned on level 1 and level 2. Level A is "appropriate" for comparing two structures of thinking, if it offers to them a finite surface for projection. This surface is equivalent to the status of recognition. The structures of thinking underlying level 1 is superior to that of level 2 because the surface. of level 1 projected onto level A is larger than the surface of level 2 projected onto level A. Level A, which is a sort of reference system, generally will have a "coarser" and "more global" structure than the two incommensurable structures of thinking (levels 1 and 2).

deviating theoretical positions. Applying this yardstick, a "later" theory must preserve and improve the practical successes of its predecessors; it is then superior to the "earlier" theories. But it must always be kept in mind that the judgement of a given structure of thinking from the pragmatic level provides only a restricted perspective.

Scientific progress defined in this way will depend in many domains essentially on the technological progress because natural science often needs a more sophisticated technology in order to perform its increasingly complicated interactions with nature. According to what has been said before, from the cognitive point of view, natural science repeatedly starts from its origin. However, given the limits imposed by technology, this process, for merely practical reasons, will proceed at an ever slower rate.

E.8 SUMMARY

Within the asymptotic convergentism it can rather be assumed that the world is actually as science envisages it to be. Here, fundamentally new pictures of the world are not generated again and again, but one frame is filled successively. In this case the concept is justified that theoretical terms like electrons, quarks, etc. are actually existing entities in the world.

However, as already outlined in detail above, the asymptotic convergentism can no longer be upheld. The empirical finding, resulting from the analysis of facts provided by the history of science, that science progresses by a sequence of incommensurable schemes of thinking (pictures) must be given a rank equal to that of a relevant experiment in a laboratory. In this connection the following point is relevant: Because a metric for the measurement of the "distance" between bodies of knowledge is not definable, there is no possibility to express certain peculiarities of the absolute truth and basic (absolute) reality, respectively.

E.9 FINAL REMARKS

We have asked the following important question: Can the progress in science, in particular in physics, lead to a final truth of the physical world? It cannot! The absolute truth, which is so to speak embedded in the basic reality, can never be observed and, therefore, theoretical considerations about it make no sense.

The perception of the complete reality in the sense of a precise reproduction implies that with growing fine structures an increasing amount of information of the outside world is needed. Then, the evolution would have furnished the sense organs with the property to transmit as much information from the outside world as possible. But the opposite is correct: The strategy of nature is to take up as little information from the reality as possible. Reality outside is not assessed by "complete" and "incomplete" but by "favourable towards life" and "hostile towards life".

The common or naive point of view assumes the following: The inside world which we feel to be outside us, actually exists in the outside world in exactly the same form as we perceive (standpoint of container principle). According to this view there is only one difference between the inside world and the outside world: inside there are only geometrical positions, whereas outside there are the real material bodies instead of the geometrical positions. In other words, it is normally assumed that the geometrical positions are merely replaced by material objects.

But why should events in nature occur, so to speak, twice, once outside of us, and again in the form of a picture? This would be against the "principles of evolution" and the "principle of usefulness", respectively. We have discussed this question in Chap. 2.

It would make not much sense if events in a world, which is tailored to fulfil the principle of usefulness, would take place twice. From the point of view of evolution the impressions before us are not precise reproductions of reality but merely appropriate pictures of it, formed by the individual from certain pieces of information from the outside world. According to the principles of evolution the central factor is "favourable towards survival" versus "hostile towards survival" (see in particular Chap. 2). The formation of a "true" picture of the world in an absolute sense, which is complete and that represents the absolute truth, is irrelevant. An individual registers situations in the environment in certain patterns which are tailor-made for the particular needs of the species and which are completely free of any compulsion towards precise "objectivity". So, a visitor of a cinema does not get at the pay desk a small but true model of the cinema (a precise reproduction of the cinema) in order to find a certain place in the cinema; a simple cinema ticket with the essential information is more appropriate. In this respect, the cinema ticket is a picture of the cinema.

The facts from the theory of science have taught us that we will never be able to make final and complete statements about the physical universe, i.e., from the point of view of the theory of science the "absolute truth" can never be obtained. Why? Because a metric for the measurement of the "distance" between bodies of knowledge is not definable; there is no possibility to express certain peculiarities of the absolute truth and basic (absolute) reality, respectively. An absolute reality may exist within the theory of science but it is not accessible to man. No doubt, the results that follow from the theory of science strongly confirm the statements, which we have worked out in connection with the projection principle.

Appendix F

BASIC ELEMENTS OF PROJECTION THEORY

F.1 HUMAN BEINGS AND OTHER BIOLOGICAL SYSTEMS

In Sec. 2.2.3 we studied a human observer *S* and an observer *S*' of another type. The material reality of *S*' should be different from the material reality of the human observer *S*. Both observers select spontaneously from the basic reality a certain part; it is *A* in the case of *S* and *A*' in the case of *S*'. The information content of the basic reality has been marked by *C*. Due to the principle of "as little outside world as possible" we have $A < C$ and $A' < C$. Since *S*' is different from *S*, we have $A' \neq A$. In other words, the information *A*' should be different from information *A*. Information *A* defines the "material reality" of *S*, and the information *A*' defines the material reality of *S*'. This in particular means that the material reality is an observer-dependent peculiarity. It is a construction (definition) by the observer, in this case *S* and *S*'.

Material reality in particular means within the projection theory that it appears as a "geometrical structure" within the brain of an observer. Since both observers, *S* and *S*', are different from each other, the frames on which the facts are projected should be different from each other. In the case of *S* we have the space-time elements x, y, z and time τ, and in the case of *S*' we mark the elements of the projection frame by $a, b, c, \ldots$ (Sec. 2.2.3). How a human observer *S* experiences the world in front of him is known, but we can at first say nothing about how the observer of other kind experiences "his" world. Sure, we can picture observer *S*' within the space and time of *S*, i.e., on the basis of the elements x, y, z and τ and, on the other hand, S' can possibly picture S on "his" projection frame with the elements $a, b, c, \ldots$. The spontaneous impressions in front of S', which appear in "his" everyday life, are different from the familiar impressions in front of a human being S.

F.2 TREATMENT OF SELECTED INFORMATION

In this Appendix we would like to point out how a human being (observer S) could describe the geometrical structures, which are embedded in space and time with the elements x, y, z and time τ. These geometrical structures represent the material reality of a human being, i.e., it is part A selected from the basic reality C:

$$C \Rightarrow A \tag{F1}$$

with

$$C > A \tag{F2}$$

The information A represents the complete material world defined by a human being. Normally, we investigate only certain parts $A_1, A_2, \ldots, A_n, \ldots$ of A.

The material reality is dependent on the type of the observer. The human observer is caught in space and time, and the geometrical structures in space and time exclusively describe material states. In Sec. 1.6.3 we have introduced "fictitious realities". The material processes can be considered as taking place in this reality, although it is a construction of the human mind. Fictitious reality is exclusively based on information A or on a part on it and, as already remarked, this information exclusively contains by definition the material aspects, i.e., material entities and physically real processes.

Since the basic reality is not accessible to a human observer, we need "fictitious realities" for the description of the images in space and time having the elements x, y, z and time τ. The physically real processes, constructed within fictitious realities, are projected onto space and time. We have introduced the concept of "fictitious reality" in Sec. 1.6.3, and we would like to repeat here the main facts: All human observers are caught in space and time. Since space and time, i.e., the elements x, y, z and time τ, cannot appear in the reality outside, we have to construct other variables when we try to construct theoretical conceptions for the material reality outside. We have marked these new variables by the letters R, S, T, Q. Because we are caught in space and time, the quantities R, S, T, Q have to be constructed on the basis of the space-time elements x, y, z and time τ.

In the concrete case the variables R, S, T, Q are expressed by the momentum $\mathbf{p}$ and the energy E (see Eq. (7)):

$$\begin{aligned} R &= p_x, \\ S &= p_y, \\ T &= p_z, \\ Q &= E \end{aligned}$$

where p_x, p_y and p_z are the usual components of the momentum $\mathbf{p}$ with $\mathbf{p} = (p_x, p_y, p_z)$. More details concerning fictitious realities are pointed out in [1].

Instead of space and time with x, y, z and time τ, fictitious reality is embedded in a space with the variables $\mathbf{p} = (p_x, p_y, p_z)$ and E, and let us mark this space $(\mathbf{p}, E)$-space.

If the function $\Psi(\mathbf{p}, E)$ characterizes the information A (or a part of it) in $(\mathbf{p}, E)$-space, the projection of the same information A onto space and time is given within the projection theory by a Fourier transform, and we get [1]

$$\begin{aligned} \Psi(\mathbf{r}, t) = {} & \frac{1}{(2\pi\hbar)^2} \int_{-\infty}^{\infty} \Psi(\mathbf{p}, E) \\ & \times \exp\left\{ i \left[\frac{\mathbf{p}}{\hbar} \cdot \mathbf{r} - \frac{E}{\hbar} t \right] \right\} dp_x dp_y dp_z dE \end{aligned} \qquad \text{(F3)}$$

where $\hbar$ is Planck's constant. In this way we get $\Psi(\mathbf{r}, t)$ from $\Psi(\mathbf{p}, E)$, and a new variable t appears in (F3) which is not known in the conventional quantum theory. The vector $\mathbf{r}$ combines the coordinates x, y, z: $\mathbf{r} = (x, y, z)$.

Essentially, within the projection theory space and time are expressed by the variables $\mathbf{r}$ and t; let us mark this space by $(\mathbf{r}, t)$-space, in analogy to $(\mathbf{p}, E)$-space.

Because the information in both spaces are exactly the same, we can apply the inverse Fourier transform

$$\Psi(\mathbf{p}, E) = \frac{1}{(2\pi\hbar)^2} \int_{-\infty}^{\infty} \Psi(\mathbf{r}, t) \times \exp\left\{-i\left[\frac{\mathbf{p}}{\hbar} \cdot \mathbf{r} - \frac{E}{\hbar} t\right]\right\} dt dx dy dz \tag{F4}$$

That is, we are able to express the information $\Psi(\mathbf{p}, E)$ of $(\mathbf{p}, E)$-space by the information $\Psi(\mathbf{r}, t)$ of $(\mathbf{r}, t)$-space. The situation has already been pointed out in Sec. 2.2.2, Fig. 23. However, we can principally not deduce from $\Psi(\mathbf{r}, t)$ or $\Psi(\mathbf{p}, E)$ the total information C of the basic reality (Sec. 2.2.2, Fig. 22). The quantities $\Psi(\mathbf{r}, t)$ and $\Psi(\mathbf{p}, E)$ will be named "wave functions".

F.3 EXTENSION OF THE TRADITIONAL VIEW

We have to conclude that there is a real extension of quantum theory when we go from the conventional quantum theory to the projection theory. Within the conventional quantum theory we have only the variable $\mathbf{r} = (x, y, z)$ and the external time parameter τ. Within the projection theory we have not only the variable $\mathbf{r} = (x, y, z)$ and τ but we have in addition the system-specific time t. In other words, when we go from the usual quantum theory to projection theory we have the transition

$$\mathbf{r}, \tau \rightarrow \mathbf{r}, t, \tau \tag{F5}$$

The additional variable t is important for the basic understanding of the notion "time".

In essence, the projection theory delivers the quantum laws in an extended form. In contrast to the usual quantum theory we have here a real symmetry between space and time; we will recognize that below.

F.4 SPECIFIC EQUATIONS

Within classical descriptions of phenomena such as, for example, Newton's equations of motion, the variables **r**, **p**, E, as well as the external time parameter τ, exist without any uncertainty.

In [1] we based our analysis on classical laws, and we used the well-known classical equation for a system (particle) in an external field $U(x, y, z)$:

$$E = \frac{\mathbf{p}^2}{2m_0} + U(x, y, z) \tag{F6}$$

where m_0 is the mass of the particle, **p** its momentum and E its energy. Here the particle is still point-like and it is embedded in space. The effect of Planck's constant $\hbar$ is that we have no longer one space, but two co-existing spaces [($\mathbf{p}$, E)-space and ($\mathbf{r}$, t)-space] representing reality and, on the other hand, its picture. Since both spaces are equivalent we may describe the physical system in ($\mathbf{p}$, E)-space as well as in ($\mathbf{r}$, t)-space. Using Eq. (F6) and the rules derived in [1], we can formulate the corresponding quantum-theoretical equations for $\Psi(\mathbf{p}, E)$ and $\Psi(\mathbf{r}, t)$.

When we apply the Fourier transforms for $\Psi(\mathbf{r}, t)$ and $\Psi(\mathbf{p}, E)$ expressed by Eqs. (F3) and (F4), we obtain for the determination of $\Psi(\mathbf{r}, t)$ and $\Psi(\mathbf{p}, E)$ the following laws:

$$i\hbar\frac{\partial}{\partial t}\Psi(\mathbf{r}, t) = -\frac{\hbar^2}{2m_0}\Delta\Psi(\mathbf{r}, t) + V(x, y, z, t)\Psi(\mathbf{r}, t) \tag{F7}$$

and

$$E\,\Psi(\mathbf{p}, E) = \frac{\mathbf{p}^2}{2m_0}\Psi(\mathbf{p}, E) + V(i\hbar\frac{\partial}{\partial p_x}, i\hbar\frac{\partial}{\partial p_y}, i\hbar\frac{\partial}{\partial p_z}, -i\hbar\frac{\partial}{\partial E})\Psi(\mathbf{p}, E), \tag{F8}$$

where $V(x, y, z, t)$ is the quantum-theoretical potential defined in [1], which takes in (F7) the form of an operator

$$V\left(i\hbar\frac{\partial}{\partial p_x}, i\hbar\frac{\partial}{\partial p_y}, i\hbar\frac{\partial}{\partial p_z} - i\hbar\frac{\partial}{\partial E}\right), \tag{F9}$$

The law (F8) determines the interaction processes in ($\mathbf{p}$, E)-space. Note that V in Eq. (F7) is different from the classical potential U (see Eq. (F6)).

The quantity

$$-i\hbar\frac{\partial}{\partial E} \tag{F10}$$

in Eq. (F8) represents the operator for the time-coordinate t. t appears as an operator when we work with the variables $\mathbf{p}$ and E of ($\mathbf{p}$, E)-space. The operator (F10) for the time-coordinate is unknown in the traditional quantum theory; here only the time τ exists, which is merely an index, i.e., it is a simple parameter and time never appears as an operator in the traditional quantum theory. The meaning of $V(x, y, z, t)$ and the corresponding operator (F9) within the projection theory will be discussed below.

The law $\Psi(\mathbf{r}, t)$ $[\Psi^*(\mathbf{r}, t)\Psi(\mathbf{r}, t)]$ is independent of the reference time τ, i.e., we have at each time τ exactly the same curve for $\Psi(\mathbf{r}, t)$ $[\Psi^*(\mathbf{r}, t)\Psi(\mathbf{r}, t)]$ (see also Fig. F1). (Ψ^* is the conjugate complex function to Ψ.)

Uncertainty relations

The variables $\mathbf{r}$, t, $\mathbf{p}$ and E are inherently uncertain and we obtain the following uncertainty relations:

$$\delta p_x \delta x \geq \tfrac{\hbar}{2},\ \delta p_y \delta y \geq \tfrac{\hbar}{2},\ \delta p_z \delta z \geq \tfrac{\hbar}{2} \tag{F11}$$

$$\delta E \delta t \geq \tfrac{\hbar}{2} \tag{F12}$$

The quantities $\delta p_x, \ldots, \delta x, \ldots, \delta E$ and δt are the uncertainties in the values x, y, z , t, p_x, p_x, p_x and E at the same instant τ. The uncertainty relations (F11) are in this form also known in usual quantum theory, but not relation (F12). As we have already pointed out in Appendix C, in the traditional quantum theory there does not exist a proper uncertainty relation for the time and the energy, and this is because a system-specific time t is not defined here.

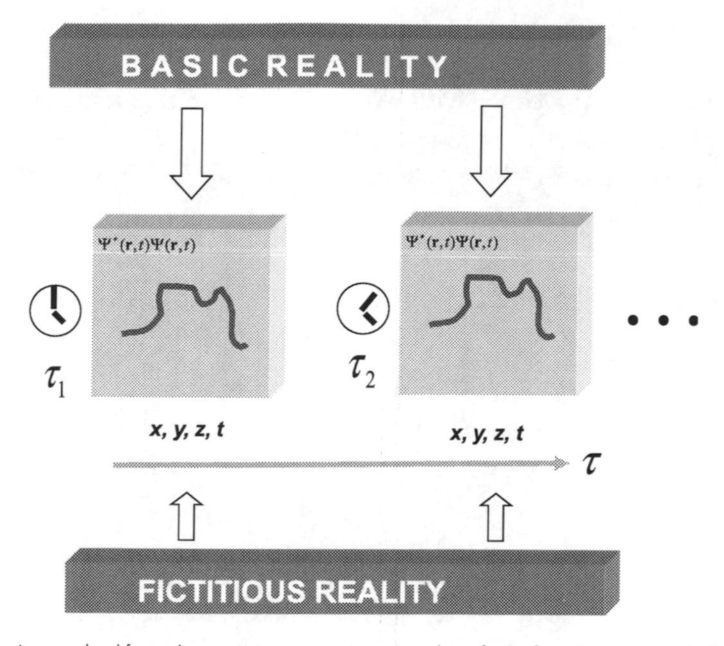

Fig. F1 In everyday life we have certain pictures spontaneously in front of us. In conventional physics it is normally assumed that real matter is embedded in space and time. Within projection theory there can only be geometrical structures in the picture and are exclusively elements (states) of the brain. All the real actions and processes, respectively, take place in absolute (basic) reality and the entire complex, which is involved, is projected onto $(\mathbf{r}, t)$-space. We describe the geometrical structures by means of fictitious realties, i.e., by $\mathbf{p}$, E-fluctuations (interactions) in. $(\mathbf{p}, E)$-space. The $\mathbf{p}$, E-structures produced by these fluctuations are projected onto $(\mathbf{r}, t)$-space and appear here as correlations. If the system under investigation is stationary with respect to time τ, the law $\Psi^*(\mathbf{r}, t)\Psi(\mathbf{r}, t)$ is independent on τ, i.e., we have at each time τ (for example, at τ_1 and τ_2) exactly the same $\mathbf{r}$, t-structure.

Stationary with respect to time t

If we assume that the potential is not dependent on time t, we have

$$V(x, y, z, t) = V(x, y, z) \tag{F13}$$

It is easy to verify that in this case the function $\Psi(\mathbf{r}, t)$ becomes zero [1]:

$$\Psi(\mathbf{r}, t) = 0 \tag{F14}$$

This result shows that the ansatz (F13) is not of relevance. More details are given in [1].

The Non-Stationary Case

The non-stationary cases within the conventional quantum theory and the projection theory refer both to the reference time τ. In the case of the projection theory the non-stationary case is expressed by [using the variables of $(\mathbf{r}, t)$-space]

$$i\hbar\frac{\partial}{\partial t}\Psi(\mathbf{r}, t)_\tau = -\frac{\hbar^2}{2m_0}\Delta\Psi(\mathbf{r}, t)_\tau + V(x, y, z, t)_\tau\Psi(\mathbf{r}, t)_\tau \qquad \text{(F15)}$$

with $\Psi_\tau \equiv \Psi(\mathbf{r}, t, \tau)$. Thus, the law $\Psi(\mathbf{r}, t)_\tau$ $[\Psi^*(\mathbf{r}, t)\Psi(\mathbf{r}, t)_\tau]$ is dependent on the reference time τ, i.e., we have at each time τ curves for $\Psi^*(\mathbf{r}, t)\Psi(\mathbf{r}, t)$ that are different from each other (see also Fig. F2).

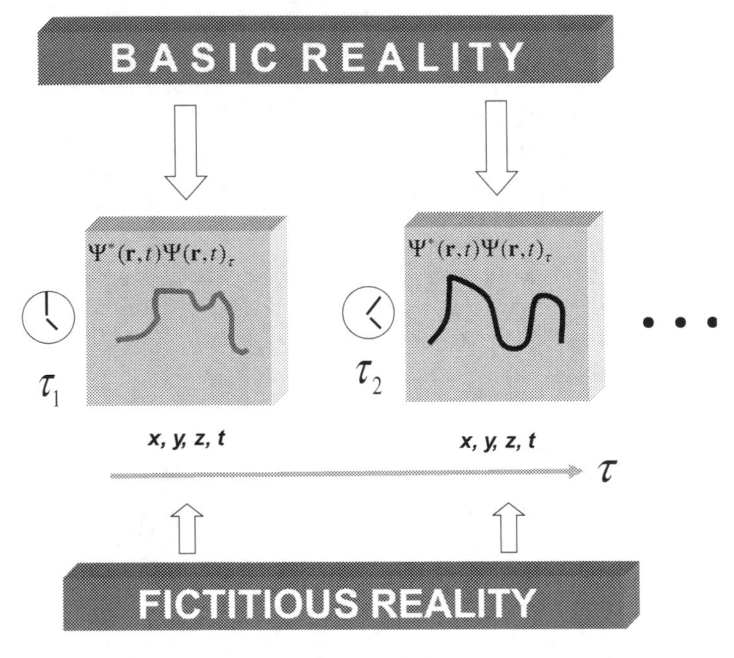

Fig. F2 In the case of τ-dependent systems (systems, which are not stationary with respect to τ) the wave function varies in the course of time τ. That is, instead of $\Psi(\mathbf{r}, t)$ we get $\Psi(\mathbf{r}, t)_\tau$. In contrast to Fig. F1, the space-time structure at τ_1 is different from that at time τ_2. As in the stationary case (Fig. F1), all the real material processes take place in basic reality and are projected onto $(\mathbf{r}, t)$-space. Also here, we describe the geometrical structures by means of fictitious realties.

The corresponding law, expressed by the variables of $(\mathbf{p}, E)$-space, has the form [1]

$$E\,\Psi(\mathbf{p}, E)_\tau = \frac{\mathbf{p}^2}{2m_0}\Psi(\mathbf{p}, E)_\tau + V\left(i\hbar\frac{\partial}{\partial p_x}, i\hbar\frac{\partial}{\partial p_y}, i\hbar\frac{\partial}{\partial p_z},\right.$$
$$\left. -i\hbar\frac{\partial}{\partial E}\right)_\tau \Psi(\mathbf{p}, E)_\tau \qquad \text{(F16)}$$

The stationary behaviour in connection with the functions $\Psi(\mathbf{r}, t)_\tau$ and $\Psi(\mathbf{p}, E)_\tau$ comes into play through the potential, that is, we obtain the stationary wave functions $\Psi(\mathbf{r}, t)$ and $\Psi(\mathbf{p}, E)$ for the replacements

$$V(x, y, z, t)_\tau \rightarrow V(x, y, z, t) \qquad \text{(F17)}$$

$$V\left(i\hbar\frac{\partial}{\partial p_x}, i\hbar\frac{\partial}{\partial p_y}, i\hbar\frac{\partial}{\partial p_z}, -i\hbar\frac{\partial}{\partial E}\right)_\tau$$
$$\rightarrow V\left(i\hbar\frac{\partial}{\partial p_x}, i\hbar\frac{\partial}{\partial p_y}, i\hbar\frac{\partial}{\partial p_z}, -i\hbar\frac{\partial}{\partial E}\right) \qquad \text{(F18)}$$

which leads to the transitions

$$\Psi(\mathbf{r}, t)_\tau \rightarrow \Psi(\mathbf{r}, t) \qquad \text{(F19)}$$

$$\Psi(\mathbf{p}, E)_\tau \rightarrow \Psi(\mathbf{p}, E) \qquad \text{(F20)}$$

Again, Eqs. (F15) and (F16) reflect non-stationary behaviour with respect to time τ, just as in the case of tradional quantum theory. As we pointed out above, the system-specific time t does not allow us to distinguish between the stationary and non-stationary classification.

The role of time

The time τ does not appear in Eq. (F3) and also not in Eq. (F4), but we obtain a new variable t for the time-coordinate. It turned out that t cannot be identified with τ. However, τ plays the role of a reference time within the projection theory and is measured by our clocks that we use for example in everyday life; the character of τ is not changed when we go from traditional physics to the projection theory.

In other words, the projection principle introduces a completely new aspect with respect to time, and time is no longer an external parameter, as in the case of τ, but t is a system-specific quantity, which characterizes, besides x, y, z the physical system under investigation. The material system produces its own t-spectrum.

Subsystems

The physical meaning of the function $\Psi(\mathbf{r}, t)$ with only two variables $\mathbf{r}$ and t is that there is only one process forming a "unified whole" without subsystems, i.e., there are no subsystems (particles) which are under mutual interaction. If there were nevertheless N interacting subsytems, having the variables

$$\mathbf{r}_1, t_1, \mathbf{r}_2, t_2, \mathbf{r}_3, t_3, \ldots, \mathbf{r}_N, t_N,$$

we would have $\Psi(\mathbf{r}_1, t_1, \mathbf{r}_2, t_2, \mathbf{r}_3, t_3, \ldots, \mathbf{r}_N, t_N)$ instead of $\Psi(\mathbf{r}, t)$:

$$\Psi(\mathbf{r}, t) \Rightarrow \Psi(\mathbf{r}_1, t_1, \mathbf{r}_2, t_2, \mathbf{r}_3, t_3, \ldots, \mathbf{r}_N, t_N) \qquad \text{(F21)}$$

This is however hardly imaginable because we somehow expect that the N subsystems are able to exist without interaction, but this is forbidden within the projection theory. The N particles can only be created if there is simultaneously an interaction between them; the N particles can not exist without such interaction processes.

Equation (F7) is based on the classical Eq. (F6). But it can be of advantage to use instead of (F6) a more general classical expression. In Eq. (F6) we have the following general relationship between the classical variables E, $\mathbf{p} = (p_x, p_y, p_z)$ and $\mathbf{r} = (x, y, z)$:

$$E = E(p_x, p_y, p_z, x, y, z)$$

As in the case of Eq. (F7), in the quantum-theoretical description, the right-hand side of this expression must always be an operator, then the energy becomes an operator when we work with the variables of $(\mathbf{r}, t)$-space. Using the operator rules derived in [1], we get the following relation:

$$i\hbar\frac{\partial}{\partial t}\Psi(\mathbf{r}, t) = \hat{H}\left(-i\hbar\frac{\partial}{\partial x}, -i\hbar\frac{\partial}{\partial y}, -i\hbar\frac{\partial}{\partial z}, x, y, z, t\right)\Psi(\mathbf{r}, t) \qquad \text{(F22)}$$

On the other hand, when we work with the variables of $(\mathbf{p}, E)$-space, we immediately obtain

$$E\,\Psi(\mathbf{p}, E) = \hat{H}\left(i\hbar\frac{\partial}{\partial p_x}, i\hbar\frac{\partial}{\partial p_y}, i\hbar\frac{\partial}{\partial p_z}, -i\hbar\frac{\partial}{\partial E}, p_x, p_y, p_z\right)\Psi(\mathbf{p}, E) \qquad \text{(F23)}$$

Equations (F22) and (F23) are more general than (F7) and (F8) and are therefore more flexible. However, the principles for the concrete determination of the laws are not known yet.

F.5 MEANING OF THE WAVE FUNCTIONS

For direct observations it is not the wave function $\Psi(\mathbf{r}, t)$ $[\Psi(\mathbf{p}, E)]$ but it is the quantity $\Psi^*(\mathbf{r}, t)\Psi(\mathbf{r}, t)$ $[\Psi^*(\mathbf{p}, E)\Psi(\mathbf{p}, E)]$ that is relevant; this point has been discussed in [1, 2]. It turned out that the quantity $\Psi^*(\mathbf{r}, t)\Psi(\mathbf{r}, t)$ $[\Psi^*(\mathbf{p}, E)\Psi(\mathbf{p}, E)]$ has to be interpreted as the probability density for the determination of the probability to find one of the values of the variables $\mathbf{p}$, E, r, t at time τ. We have analyzed the situation in [1] in more detail. Let us repeat here only some basic facts.

Since the variables $\mathbf{r}, t$ are only existent in connection with real bodies (see in particular Chap. 1), the following question arises: What is the probability density $\Psi^*(\mathbf{r}, t)\Psi(\mathbf{r}, t)$? From Eq. (F3) it directly follows that $\Psi(\mathbf{r}, t)$ is determined at location $(\mathbf{r}, t)$ in space-time by all possible values of $\mathbf{p}$ and E $(-\infty < \mathbf{p},\ E < \infty)$ which are given with the probability density $\Psi^*(\mathbf{p}, E)\Psi(\mathbf{p}, E)$. Therefore, $\Psi^*(\mathbf{r}, t)\Psi(\mathbf{r}, t)$ can only be interpreted in connection with the variables $\mathbf{p}$ and E, which reflect the material reality. There is no other way: *One of the possible* *values for* $\mathbf{p}$ *and for* E *is present in the intervals* $\mathbf{r}, \mathbf{r} + d\mathbf{r}$ *and* $t, t + dt$ *with the probability density of* $\Psi^*(\mathbf{r}, t)\Psi(\mathbf{r}, t)$.

Since only the variables $\mathbf{p}$ and E are accessible to measurements we can also state: *The measurement of one of the possible values for* p *and for* E *is done in the space-time intervals* $\mathbf{r}, \mathbf{r} + d\mathbf{r}$ *and* $t, t + dt$ *with the probability density of* $\Psi^*(\mathbf{r}, t)\Psi(\mathbf{r}, t)$.

This is actually the situation in practical experiments. Signals are recorded with detectors in space and time, no more, no less. This has primarily nothing to do with a particle localized in space.

In other words, at time τ we have probability distributions $\{\mathbf{r}\}$, $\{t\}$, $\{\mathbf{p}\}$, $\{E\}$ for the variables $\mathbf{p}$, E, r, t:

$$\tau : \{\mathbf{r}\}, \{t\}, \{\mathbf{p}\}, \{E\} \tag{F24}$$

Only *one* value of each distribution can be realized at time τ, and there is a certain probability for the existence of these values which are expressed by $\Psi^*(\mathbf{r}, t)\Psi(\mathbf{r}, t)$ and $\Psi^*(\mathbf{p}, E)\Psi(\mathbf{p}, E)$.

While the spaces [$(\mathbf{r}, t)$-space and $(\mathbf{p}, E)$-space] can be thought of as infinite, the ranges of $\mathbf{p} = (p_x, p_y, p_z)$, E, $\mathbf{r} = (x, y, z)$, t of real material systems can be arbitrary but should be different from infinity [1, 2], i.e., in general we have

$$\begin{aligned} \tau : -\infty < x, y, z < \infty \\ -\infty < t < \infty \end{aligned} \tag{F25}$$

and

$$\begin{aligned} \tau : -\infty < p_x, p_y, p_z < \infty \\ -\infty < E < \infty \end{aligned} \tag{F26}$$

Clearly, the ranges of the variables are defined by the ranges for which the probability densities $\Psi^*(\mathbf{r}, t)\Psi(\mathbf{r}, t)$ and $\Psi^*(\mathbf{p}, E)\Psi(\mathbf{p}, E)$ are not zero. One of these values, defined within these ranges, are given at time τ with a certain probability. Note that also the system-specific time t is at any time τ expressed (observed) by a probability distribution.

Let us illustrate the situation in somewhat more detail. The observer measures with his clock at certain times $\tau_1, \tau_2, \cdots, \tau_i, \cdots$

certain values for $p_x, p_y, p_z, E, x, y, z, t$, and he gets

$$\begin{aligned}
\tau_1 &: \; p_{x1}, p_{y1}, p_{z1}, E_1, x_1, y_1, z_1, t_1 \\
&\quad (-\infty < p_{x1}, p_{y1}, p_{z1}, E_1, x_1, y_1, z_1, t_1 < \infty) \\
\tau_2 &: \; p_{x2}, p_{y2}, p_{z2}, E_2, x_2, y_2, z, t_2 \\
&\quad (-\infty < p_{x2}, p_{y2}, p_{z2}, E_2, x_2, y_2, z_2, t_2 < \infty) \\
&\quad \vdots \\
\tau_i &: \; p_{xi}, p_{yi}, p_{zi}, E_i, x_i, y_i, z_i, t_2 \\
&\quad (-\infty < p_{xi}, p_{yi}, p_{zi}, E_i, x_i, y_i, z_i, t_i < \infty) \\
&\quad \vdots
\end{aligned} \tag{F27}$$

Here the probability densities $\Psi^*(\mathbf{r}, t)\Psi(\mathbf{r}, t)$ and $\Psi^*(\mathbf{p}, E)\Psi(\mathbf{p}, E)$ are assumed not to be dependent on τ.

The aspect of t is of particular interest, and we would like to repeat what we have pointed out in [2]: Whereas the reference time τ goes strictly from the past to the future, the system-specific time t jumps in general arbitrarily from one t-position to another, and we cannot know when we observe — with respect to time τ — an event in connection with the system under investigation in the past, present or in the future. The following example illustrates that:

When we measure at time τ_i one space-time point of the configuration $\Psi^*(\mathbf{r}, t_i)\Psi(\mathbf{r}, t_i)$, then the result of such measurement could be given for example by

$$\begin{aligned}
\tau_1 &\rightarrow t_2 \\
\tau_2 &\rightarrow t_4 \\
\tau_3 &\rightarrow t_1 \\
\tau_4 &\rightarrow t_3
\end{aligned} \tag{F28}$$

in the case of $i = 1, \ldots, 4$, that is, the sequence with respect to the reference time is given by

$$t_2, t_4, t_1, t_3 \tag{F29}$$

for

$$\tau_1 < \tau_2 < \tau_3 < \tau_4 \tag{F30}$$

and

$$t_1 < t_2 < t_3 < t_4 \tag{F31}$$

Because of the statistical behaviour of the system-specific time t the same measurement could have led also to the following result:

$$\begin{aligned} \tau_1 &\rightarrow t_3 \\ \tau_2 &\rightarrow t_1 \\ \tau_3 &\rightarrow t_2 \\ \tau_4 &\rightarrow t_4 \end{aligned} \tag{F32}$$

In this case we have the sequence

$$t_3, t_1, t_2, t_4 \tag{F33}$$

which is different from (F29). Clearly, here (F30) and (F31) are valid.

In general, the system-specific time t jumps arbitrarily from one t-position to another, and we cannot know when we observe an event in the past, present or in the future. The reference time τ, measured with our clocks, goes by definition always strictly from the past to the future.

This feature of time is completely new. Even the notion "system-specific time" t is not known in conventional quantum theory; here only the reference time τ appears which however only plays the role of an external parameter, as we have already mentioned above.

F.6 COUNTERPARTS

The system under investigation, say i, is described by the wave functions $\Psi(\mathbf{r}, t)$ and $\Psi(\mathbf{p}, E)$. For the probability of finding the variables $\mathbf{r}$ and t in $(\mathbf{r}, t)$-space the quantity $\Psi^*(\mathbf{r}, t)\Psi(\mathbf{r}, t)$ is relevant. On the other hand, the occurrence of the values of the variables $\mathbf{p}$ and E in $(\mathbf{p}, E)$-space is characterized by $\Psi^*(\mathbf{p}, E)\Psi(\mathbf{p}, E)$. The quantity $\Psi^*(\mathbf{r}, t)\Psi(\mathbf{r}, t)$ defines the system (particle) in $(\mathbf{r}, t)$-space, that is, its geometrical form and structure, respectively.

F.6.1 Conservation Laws

The point is that the momentum and the energy of system i fluctuate incessantly because we have at each time τ other values for $\mathbf{p}$ and E. If the values for the momentum and the energy are at time τ_1 given by $p_{x1}, p_{y1}, p_{z1}, E_1$ and those at τ_2 by $p_{x2}, p_{y2}, p_{z2}, E_2$ (see Eq. (F27)), where τ_2 is assumed to be the next possible value after τ_1, we have the $\mathbf{p}, E$-fluctuations $\Delta\mathbf{p} = (\Delta p_x, \Delta p_y, \Delta p_z)$ and ΔE.

$$\begin{aligned}
\Delta p_x &= p_{x2} - p_{x1} \\
\Delta p_y &= p_{y2} - p_{y1} \\
\Delta p_z &= p_{z2} - p_{z1} \\
\Delta E &= E_2 - E_1
\end{aligned}$$

Since the $\mathbf{p}, E$-values fluctuate for the system i (in principal in the range $-\infty < \mathbf{p}, E < \infty$) there must be another system, say j, a counterpart of the system i, in order to fulfil the conservation laws for momentum and energy: System i and system j exchange momentum and energy in accordance with the conservation laws, i.e., we must have

$$\begin{aligned}
\mathbf{p}_i(\tau_2) &= \mathbf{p}_i(\tau_1) \pm \Delta\mathbf{p} \\
E_i(\tau_2) &= E_i(\tau_1) \pm \Delta E \\
\mathbf{p}_j(\tau_2) &= \mathbf{p}_j(\tau_1) \mp \Delta\mathbf{p} \\
E_j(\tau_2) &= E_j(\tau_1) \mp \Delta E.
\end{aligned} \tag{F34}$$

The laws (F34) are independent of the positions of both systems (particles); the conservation laws must be fulfilled independent of the relative positions of systems i and j. System (particle) i cannot exist without the existence of system (particle) j and vice versa. Evidently, we need at least two systems; the conservation laws do demand that. The "existence-inducing interactions" are produced by $\mathbf{p}, E$-fluctuations between system i and systems j and lead to distance-independent correlations in $(\mathbf{r}, t)$-space. This kind of interaction produces the system (particle) itself, and the quantity $\Psi^*(\mathbf{r}, t)\Psi(\mathbf{r}, t)$ describes its form and shape, respectively; it can therefore be called form-interaction. We showed in [1, 2] that distance-dependent correlations can be introduced within the framework of projection principle.

Remark

It is important to note that within conventional physics only distance-dependent interactions are known, that is, form-interactions are not defined here. This is a very principal point, in particular with respect to the notion of "interaction". The source for the notion of "interaction" is an invention of classical physics and goes back to Newton. In [1, 2] we discussed the main facts in connection with distance-dependent interactions as they are defined and used in conventional physics.

F.6.2 Accordance with Basic Observations

In Sec. 1.4.6 we had to introduce distance-independent "existence-inducing interactions", and this is because existence-inducing interactions were needed to fulfil basic observation facts: The basic elements of space and time, characterized by x, y, z and τ, are not accessible to our senses. We definitely cannot see, hear, smell, or taste them. Measuring instruments for the experimental determination of the space-time points x, y, z, and τ are not known and even unthinkable. At time τ we can only say something about "distances in connection with masses", and "time intervals in connection with physical processes". In a nutshell, we need at least two systems to fulfil these basic observation laws.

Projection theory exactly fulfils these basic observation facts. The theory automatically leads to the result that lone systems cannot exist, but we need at least two systems (particles). The conservation laws demand this condition. Projection theory seems to be adapted to the basic observation facts (at time τ we can only say something about "distances in connection with masses", and "time intervals in connection with physical processes"). However, here real matter is not embedded in space and time, but we have merely geometrical structures in $(\mathbf{r}, t)$-space. Nevertheless, our basic statement "at time τ we can only say something about *distances in connection with masses*, and *time intervals in connection with physical processes*", formulated above and in Sec. 1.4.1, is fulfilled when we replace the statement "*at time τ we can only say something about distances in connection with masses*" by the equivalent formulation "*at time τ we can only say something about distances in connection with geometrical positions*".

F.6.3 Conclusion

These $\mathbf{p}$, E-fluctuations between system i and system j define an interaction, and the $\mathbf{p}$, E-values of the two interacting systems are strongly correlated at each time τ. Only in this way the conservation laws for momentum and energy can be fulfilled. If system i would perform $\mathbf{p}$, E-fluctuations independently from system j and vice versa, these conservation laws would be violated.

Therefore, the systematic $\mathbf{p}$, E-fluctuations in $(\mathbf{p}, E)$-space reflect real physical processes within the material world. Thus, we have to consider the $(\mathbf{p}, E)$-space as reality. However, because the variables $\mathbf{p}$ and E are auxiliary elements, this reality is a "fictitious reality". Therefore, the $\mathbf{p}$, E-fluctuations (the interactions of the systems with other systems) are fictitious in character and merely simulate the real processes in the basic reality.

Again, due to the conservation laws for the momentum $\mathbf{p}$ and the energy E we must have systematic interaction processes ($\mathbf{p}$, E-fluctuations) between the system i and system j and, therefore, we have to consider the $(\mathbf{p}, E)$-space as (fictitious) reality. However, there do not exist such conservation laws for the variables $\mathbf{r}$ and t, i.e., the realized $\mathbf{r}$, t-point at time τ of system i is independent of the realized $\mathbf{r}$, t-points in connection with system j at the same time τ. There is no exchange of certain "space-time pieces" between the two pictures (see also Fig. 6). In other words, there are no correlations between the $\mathbf{r}$, t-points at time τ. This indicates that the variables $\mathbf{r}$ and t are in fact elements of a picture. In other words, the information of fictitious reality [$(\mathbf{p}, E)$-space] is projected onto $(\mathbf{r}, t)$-space by a Fourier transform and we get the "picture of reality". The situation is summarized in Fig. F3.

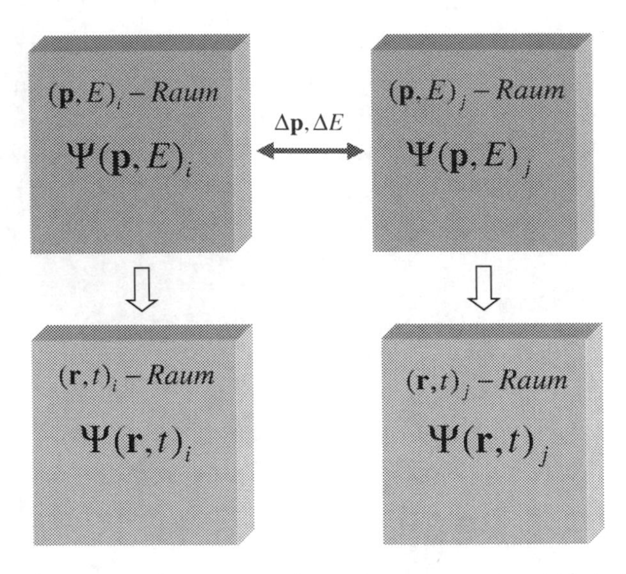

Fig. F3 Due to the conservation for momentum and energy there are systematic $\mathbf{p}$, E-fluctuations $\Delta\mathbf{p}$, ΔE between the two $(\mathbf{p}, E)_k$-spaces, $k = i, j$. However, there are no fluctuations of certain "space-time elements" between the pictures for system i and system j. Both systems are assumed to be identical.

F.7 ONLY PROCESSES ARE RELEVANT!

F.7.1 Non-interacting Systems

It can be shown within the projection theory that free (non-interacting) systems cannot exist if the system is elementary in character [1, 2], i.e., if it has no internal structure. Elementary in character means in particular that the system cannot be divided into subsystems. A system can be considered as free if the potential function becomes zero: $V(x, y, z, t) = 0$; then, the operator $V(i\hbar\partial/\partial p_x, \ldots, -i\hbar\partial/\partial E)$ (Sec. F4) is not definable.

In the case of free systems the momentum $\mathbf{p} = \mathbf{p}_0$ and the energy $E = E_0$ remain constant in the course of time τ. There is no interaction between the system under investigation and other systems (environment), and there are no $\mathbf{p}$, E-fluctuations in the case of free systems: $\Delta\mathbf{p} = 0$, $\Delta E = 0$. What do we get within the projection theory for the wave functions $\Psi(\mathbf{r}, t)$ and $\Psi(\mathbf{p}_0, E_0)$ in this specific case?

The outcome can be summarized as follows [1]:

1. The wave functions $\Psi(\mathbf{r}, t)$ and $\Psi(\mathbf{p}_0, E_0)$ become zero:

$$\Psi(\mathbf{r}, t) = 0 \tag{F35}$$

$$\Psi(\mathbf{p}_0, E_0) = 0 \tag{F36}$$

Thus, the probability densities are zero:

$$\Psi^*(\mathbf{r}, t)\Psi(\mathbf{r}, t) = 0 \tag{F37}$$

$$\Psi^*(\mathbf{p}_0, E_0)\Psi(\mathbf{p}_0, E_0) = 0 \tag{F38}$$

2. The analysis showed [1] that the system's momentum $\mathbf{p}_0$ and its energy E_0 become zero too: $\mathbf{p}_0 = 0$, $E_0 = 0$. That is, nothing has been taken away from the cosmic $\mathbf{p}, E$-pool. Therefore, from the point of view of the projection theory free elementary systems cannot exist in nature. In other words, lone systems without an inner structure cannot exist within the framework of the projection theory.

It is astonishing that this result is in accordance with what we have deduced in connection with the basic observation facts (Chap. 1): We definitely cannot observe at time τ single elements x, y, z of space. Furthermore, it is outlined in Chap. 1 that a single, isolated elementary body cannot exist and is therefore not a physically real entity. All that is in accordance with the corresponding results of the projection theory; it particularly shows that this theory is adapted to real situations.

F.7.2 Principle of Usefulness

In essence, the projection theory requires that non-interacting systems cannot exist in nature. Why does nature not admit such systems, i.e., free elementary systems without an internal structure? Such free systems with constant momentum $\mathbf{p}_0$ and constant energy E_0 are in a certain sense "dead systems" because they are not involved in the processes in nature. There are no $\mathbf{p}, E$-fluctuations between a free system and its surroundings, and such systems would be "useless". In other words, systems with

$$\mathbf{p}_0 = const \neq 0 \quad \text{(F39)}$$

$$E_0 = const \neq 0 \quad \text{(F40)}$$

have no place in nature because they would be completely detached from the scenario and this is against the "principle of usefulness", which reflects the laws of evolution. Thus, from this point of view free systems should not exist in nature. In fact, the projection theory excludes such states explicitly because we have $\mathbf{p}_0 = 0$ and $E_0 = 0$. In particular, we have $\Psi(\mathbf{r}, t) = 0$ and $\Psi(\mathbf{p}, E) = \Psi(\mathbf{p}_0, E_0) = 0$, i.e., the properties given by the Eqs. (F39) and (F40) do not exist within the projection theory. In other words, the projection theory supports the existence of the principle of usefulness.

No doubt, the world of man is formed in accordance with the principle of usefulness, and a lot of examples, in particular in connection with evolution, support this view. It is astonishing that this principle is obviously even reflected at the basic level of the theoretical description (projection theory). However, this is not a description of the "objective reality" but it is "merely" man's view; man is principally not able to recognize the properties of what we have called the "objective (basic) reality".

F.8 BLOCK UNIVERSE

F.8.1 View within the Special Theory of Relativity

The four-dimensional space-time of the Special Theory of Relativity reflects a certain kind of "block universe". Within this view the totality of material existence with respect to the past, present and future "is laid out frozen before us" [27]. Here nothing unexpected can ever happen. The situation within the Special Theory of Relativity is well summarized by Jim Al-Khalili [27]:

Minkowski's 4D spacetime is often referred to as the block universe model. Once time is treated like a fourth dimension of space we can imagine the whole of space and time modelled as a four-dimensional block . . . Here we have a view of the totality of existence in which the whole of time — past, present and future — is laid out frozen before us. Many physicists, including Einstein later in his life, pushed this model to its logical conclusion: in 4D spacetime, nothing ever moves. All events which have ever happened or ever will happen exist together in the block universe and there is no distinction between past and future. This implies that nothing unexpected can ever happen. Not only is the future preordained but it is already out there and is as unalterably fixed as the past.

Is this picture really necessary? After all, we can just as easily imagine a Newtonian spacetime modelled as a 4D block. The difference is that in that case space and time are independent of each other, whereas in relativity the two are linked. One of the consequences of relativity is that no two observers will be able to agree on when 'now' is. By abandoning absolute time we must also admit that the notion of a universal present moment does not exist. For one observer, all events in the Universe that appear to be simultaneous can be linked together to form a certain cross sectional slice through spacetime which that observer calls 'now'. But another observer, moving relative to the first, will have a different slice that will cross the first. Some events that lie on the first observer's 'now' slice will be in the second observer's past while others will be in his future. This mind-boggling result is known as the relativity of simultaneity, and is the reason why many physicists have argued that since there is no absolute division between past and future then there can be no passage of time, since we cannot agree on where the present should be.

Worse than that, if one observer sees an event A occur before an event B, then it is possible for another observer to witness B before A. If two observes cannot even agree on the order that things happen, how can we ever define an objective passage of time as a sequence of events?" [27]

In short, within the Special Theory of Relativity we have a specific block universe consisting of a four-dimensional space-time. The totality of information in connection with the past, present and future "is laid out frozen before us". Question: Is everything within this block universe, i.e., within this (x, y, z, τ)-block, only "determined" or is it really "existent at one go"? (Note that the system-specific time t is not defined within conventional physics.) What does "determined" and "existent at one go" here mean?

1. If the (x, y, z, τ)-block is only "determined", the material world only exists at a certain time, say τ_i. The material world is annihilated at time $\tau_{i-1} < \tau_i$ when it is existent at τ_i. In other words, the material reality is incessantly created and annihilated in the course of time τ, but it is entirely determined for all times τ. However, since an "absolute present" is not definable in the Special Theory of Relativity this case should be excluded.
2. If the material world is "existent at one go", i.e., for all times τ, it is laid out frozen before us. It becomes conscious to the observer at time $\tau = \tau_i$. It is however unconscious to the observer for times $\tau \neq \tau_i$.

(Within Newton's mechanics we have a three-dimensional block, i.e., we have a (x, y, z)-block. The world (system) only exists in the present, but not in the past and the future. Here the world (system) moves from the past over the present to the future. If the time τ_P defines the present, the world (system) exists for $\tau = \tau_P$ but not for $\tau < \tau_P$ and $\tau > \tau_P$. This behaviour is directly coupled to consciousness; it is perceived at $\tau = \tau_P$, but not for $\tau < \tau_P$ and $\tau > \tau_P$.

To sum up, the block universe of the Special Theory of Relativity reflects in any case a static reality, but certain questions remain: Does the entire material world "exist" as a whole or is everything "determined" for ever? In the case where consciousness comes into play, we have to find out how consciousness is coupled to the block universe defined by the Special Theory of Relativity. However, all these questions possibly become superfluous because this block universe contains matter and, therefore, it is organized after the "container principle" (Chap. 1) that we have excluded for principal reasons.

F.8.2 View within the Projection Theory

All the physical developments in conventional physics (Newton's mechanics, the Theory of Relativity, the usual quantum theory, etc,) are based on the assumption that we live in "one" world, which exists independently from the observer, i.e., it is assumed that the material reality exists objectively. It is also believed that the human observer is able to seize the complete information of this kind of universe. Therefore, most scientists are firmly convinced to soon have the absolute truth in their hands and are seduced to create what is often called the "world equation". In fact, the "container principle" suggests this view, which however has to be seen as a questionable conception (Chap. 1).

Within the projection theory we have in principal not only "one" material reality, but as many realities as there are different biological systems. Each species defines its own "material world"; the details have been pointed out in Sec. 2.2.3. Other biological systems experience a material world that is different from that of human beings, at least in principle. All is dependent on the information that an individual selects from the basic reality. Thus, the term "world equation" is not applicable here. Each species has its own "world equation", which however can only reflect a certain part of the basic reality.

F.8.2.1 *Number of events*

At time τ the whole $\mathbf{r}, t$-scenario is given:

$$\Psi^*(\mathbf{r}, t)\Psi(\mathbf{r}, t), \quad -\infty < \mathbf{r} < \infty, \quad -\infty < t < \infty \qquad \text{(F41)}$$

In other words, at time τ one of the values of $\mathbf{r} = (x, y, y, z)$ and t, for which $\Psi^*(\mathbf{r}, t)\Psi(\mathbf{r}, t)$ is not zero ($\Psi^*(\mathbf{r}, t)\Psi(\mathbf{r}, t) \neq 0$), are defined with a certain probability. This is valid for all times τ. Equation (F41) defines a block universe within the projection theory; it describes the selected part A (or a part of it) of the total information C of the basic reality (for details, see Sec. 2.2.3). Note in particular that at a certain time τ the past, present and future with respect to the system-specific time t can be observed.

Furthermore, we have relevant properties here: At time τ we have probability distributions $\{\mathbf{r}\}$, $\{t\}$, $\{\mathbf{p}\}$, $\{E\}$ for the variables r, t, $\mathbf{p}$ and E: (Eq. (F24). However, only "one" value of each distribution can be realized at time τ, and there is a certain probability for the existence of these values which is expressed by $\Psi^*(\mathbf{r}, t)\Psi(\mathbf{r}, t)$ and $\Psi^*(\mathbf{p}, E)\Psi(\mathbf{p}, E)$. But how many values of each variable $\mathbf{r}$, t, $\mathbf{p}$ and E could in principle be registered within a small time interval $\Delta\tau$ different from zero? This can be easily estimated by means of some simple arguments.

Let $\Delta_\mathbf{r}$, Δ_t, $\Delta_\mathbf{p}$, Δ_E be the ranges of $\mathbf{r}$, t, $\mathbf{p}$ and E for which the distributions $\Psi^*(\mathbf{r}, t)\Psi(\mathbf{r}, t)$ and $\Psi^*(\mathbf{p}, E)\Psi(\mathbf{p}, E)$ are not zero, and let us assume that the ranges $\Delta_\mathbf{r}$, Δ_t, $\Delta_\mathbf{p}$, Δ_E are different from infinity. How many values $\mathbf{r}$, t, $\mathbf{p}$ and E can come into existence in the time interval $\Delta\tau = \varepsilon$, where ε is infinitesimal but different from zero? This number of values for each variable is identical with the number of τ-values in the interval ε, and this number is given by the number of all real numbers within ε. We know from mathematics that the number of real numbers in ε must be infinity, and in physics none of these real numbers are excluded. Therefore, within the infinitesimal as well, interval ε the number N of values of a variable (as, for example $\mathbf{r}$) is infinite in the interval $\Delta_\mathbf{r}$. Thus, the number density $\rho = N/\Delta_\mathbf{r}$ of $\mathbf{r}$-values is infinite in the interval $\Delta_\mathbf{r}$, where the interval $\Delta_\mathbf{r}$ can take any value but must be different from infinity.

Clearly, the possible values defined by $\Psi^*(\mathbf{r}, t)\Psi(\mathbf{r}, t)$ and $\Psi^*(\mathbf{p}, E)$ $\Psi(\mathbf{p}, E)$ are realized with an infinite number density already within an infinitesimal time-interval ε of almost zero, although the ranges $\Delta_\mathbf{r}$, Δ_t, $\Delta_\mathbf{p}$, Δ_E can take any value (however, it must be different from infinity). This is particularly remarkable in connection with the system-specific time t and will be analyzed in more detail below.

The block universe of projection theory is incessantly created and annihilated when we progress with respect to time τ in steps of time-intervals ε.

The fact that all possible values defined by $\Psi^*(\mathbf{r}, t)\Psi(\mathbf{r}, t)$ and $\Psi^*(\mathbf{p}, E)\Psi(\mathbf{p}, E)$ are already realized within an infinitesimal time-interval ε of almost zero is remarkable, and let us call it the ε-property. But also in classical mechanics we work with an infinite number of space-points within an infinitesimal interval $\Delta\tau = \varepsilon$. Let us consider a classical particle, which moves with velocity v_x from one space-position to another defining the space-distance of $\Delta_{\mathbf{r}}$. If this process takes place within an infinitesimal time interval $\Delta\tau = \varepsilon$, the space-interval $\Delta_{\mathbf{r}}$ must be infinitesimal as well, otherwise we will not be able to define a reasonable classical velocity v_x (classical mechanics is a local theory). Nevertheless, the interval $\Delta_{\mathbf{r}}$ contains an infinite number of real numbers, that is, the particle runs monotonically over an infinite number of real numbers within the infinitesimal time intervals $\Delta\tau = \varepsilon$.

The world is however non-local in quantum theory, and the laws are non-deterministic. In particular, the definition of the velocity $\mathbf{v}$ in the classical sense of the word is not possible. In the conventional quantum theory only the space variable $\mathbf{r}$ behaves statistically. In contrast to classical mechanics, within the projection theory the space-time intervals $\Delta_{\mathbf{r}}$ and Δ_t can take any values (different from infinity) in the case of an infinitesimal time interval $\Delta\tau = \varepsilon$. Both intervals ($\Delta_{\mathbf{r}}$ and Δ_t) are statistically occupied in the course of time τ. Within the infinitesimal time interval $\Delta\tau = \varepsilon$ the variables of the system jump statistically from one space-time point $(\mathbf{r}_i, t_i)$ to another point $(\mathbf{r}_j, t_j)$ where $\mathbf{r}_{i\,j} = \mathbf{r}_i - \mathbf{r}_j$ and $t_{i\,j} = t_i - t_j$ can take arbitrary values within the space intervals $\Delta_{\mathbf{r}}, \Delta_t$: $\mathbf{r}_{i\,j} \leq \Delta_{\mathbf{r}}$, $t_{i\,j} \leq \Delta_t$. Although the time interval $\Delta\tau = \varepsilon$ is infinitesimal, the number density in both intervals is infinity because the number of events within $\Delta\tau = \varepsilon$ is infinite. Note, that this property is independent on $\Delta_{\mathbf{r}}, \Delta_t, \Delta_{\mathbf{p}}, \Delta_E$. Again, the ranges $\Delta_{\mathbf{r}}, \Delta_t, \Delta_{\mathbf{p}}, \Delta_E$ can take any value but have to be different from infinity.

F.8.2.2 *Features*

What are the differences between the block unverse of the projection theory and the block universe of the Special Theory of Relativity? We remarked above that in the Special Theory of Relativity there is no absolute division between the past and future, and we cannot agree on where the present should be. This behaviour is also reflected in

$\Psi^*(\mathbf{r}, t)\Psi(\mathbf{r}, t)$ because the range of the time interval Δ_t is statistically occupied in the course of time τ and we cannot agree on where the present should be, just as in the case of the block universe of the Special Theory of Relativity. Without a reference time structure τ we would not be able to distinguish between the past, present and future.

The statement "in 4D space-time nothing ever moves" is in accordance with what we have pointed out above in connection with the projection theory: $\Psi^*(\mathbf{r}, t)\Psi(\mathbf{r}, t)$ is a static function and does not change; the general Eq. (F7) just describes this situation. However, in principle $\Psi^*(\mathbf{r}, t)\Psi(\mathbf{r}, t)$ could also vary with time τ (Sec. F.4). Furthermore, the statement "Not only is the future preordained but is already out there and is as unalterably fixed in the past" is also in entire accordance with that we have outlined above about the nature of time.

There is however an essential difference between the result of the Special Theory of Relativity and the investigation given here. Whereas the block universe of the Special Theory of Relativity is valid for the whole universe and cannot be influenced by the observer, the quantity $\Psi^*(\mathbf{r}, t)\Psi(\mathbf{r}, t)$ of the projection theory reflects only a certain system and can definitely influenced by an observer or by any external system. For example, the system could be put at time τ_I into another environment. Such an influence means that the interaction between the system and its environment is changed and this is described by a new wave function, say $\Omega(\mathbf{r}, t)$. In other words, instead of $\Psi^*(\mathbf{r}, t)\Psi(\mathbf{r}, t)$ we have now $\Omega^*(\mathbf{r}, t)\Omega(\mathbf{r}, t)$ for times $\tau > \tau_I$.

As in the case of $\Psi^*(\mathbf{r}, t)\Psi(\mathbf{r}, t)$, also in connection with $\Omega^*(\mathbf{r}, t)$ $\Omega(\mathbf{r}, t)$ the following is valid: For certain times $\tau > \tau_I$ only one value of the system-specific time t is realized with a certain probability. However, if we consider an infinitesimal time interval $\Delta\tau = \varepsilon$ an infinite number of t-values are occupied. In other words, the whole history (the complete past and future) of the system, described by the range Δ_t (life-time) of the distribution $\Omega^*(\mathbf{r}, t)\Omega(\mathbf{r}, t)$, is given within the infinitesimal time interval $\Delta\tau = \varepsilon$ measured by our clocks. This is the case for any $\tau_i \pm \Delta\tau$, i.e., the law $\Omega^*(\mathbf{r}, t)\Omega(\mathbf{r}, t)$ is independent of $\tau > \tau_I$; it is stationary with respect to time $\tau > \tau_I$ (see also Sec. F.4).

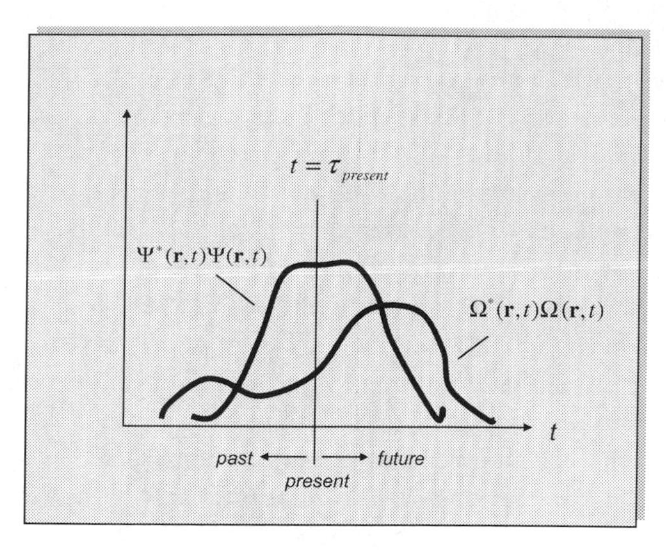

Fig. F4 The worlds, characterized by $\Psi^*(\mathbf{r}, t)\Psi(\mathbf{r}, t)$ and $\Omega^*(\mathbf{r}, t)\Omega(\mathbf{r}, t)$, are not only different at the present, characterized here by a certain time $t = \tau_{present}$, but the system is changed within the entire time scale t.

Thus, here we may give an essential statement: Despite the statistical fluctuations, the whole of time t — its past, present and future — is laid out frozen before us. But the past and the future of $\Omega^*(\mathbf{r}, t)\Omega(\mathbf{r}, t)$ have in general nothing to do the past and the future of $\Psi^*(\mathbf{r}, t)\Psi(\mathbf{r}, t)$. The worlds, characterized by $\Psi^*(\mathbf{r}, t)\Psi(\mathbf{r}, t)$ and $\Omega^*(\mathbf{r}, t)\Omega(\mathbf{r}, t)$, are not only different at the present, characterized here by a certain time $t = \tau_{present}$, but the system is changed within the entire time scale t. This is demonstrated in Fig. F4.

F.9 STRUCTURE OF REFERENCE TIME

Concerning time t we have a remarkable situation: At time τ only one value of the system-specific time t is realized with a certain probability. However, if we consider an infinitesimal time interval $\Delta\tau = \varepsilon$ an infinite number of t-values are occupied, that is, the whole history (the complete past, present and future) of the system, described by the range Δ_t (life-time) of the distribution $\Psi^*(\mathbf{r}, t)\Psi(\mathbf{r}, t)$, is given within the infinitesimal time interval $\Delta\tau = \varepsilon$ of our clocks (Section F.8).

This is the case for any $\tau_i \pm \Delta\tau$, i.e., the law $\Psi^*(\mathbf{r}, t)\Psi(\mathbf{r}, t)$ is independent of τ; it is stationary with respect to time τ (Sec. F.4, Fig. F1). In other words, despite the statistical fluctuations, the whole of time — past, present and future — is laid out frozen before us. The system fluctuates with respect to time without any direction; this feature has already been discussed in Sec. F.5.

In other words, there is no connection between τ and the system-specific time t, i.e., the introduction of τ within the projection theory, outlined so far, would make no sense. Therefore, we have to construct a connection between τ and the system-specific time t, because the existence of τ is a matter of fact. This connection between the reference time τ and the system-specific time t can be constructed as follows:

We know from our observations that we always observe only certain configurations of reality at time τ:

$$\Psi^*(\mathbf{r}, t_0)\Psi(\mathbf{r}, t_0), \quad -\infty < \mathbf{r} < \infty \tag{F42}$$

for $t_0 = \tau$. Each photography (Fig. 1) represents such a configuration in space, at a certain time $\tau = t_0$. Why does nature work in this way? Why such a selection processes from $\Psi^*(\mathbf{r}, t)\Psi(\mathbf{r}, t)$ to $\Psi^*(\mathbf{r}, t_0)\Psi(\mathbf{r}, t_0)$ (see Eqs. (F41) and (F42))? The answer is probably given by evolution. There is an important basic principle in connection with evolution [Chap. 2]; it is the principle "as little outside world as possible". This principle obviously guarantees optimal chances for survival and is clearly reflected in the transition from $\Psi^*(\mathbf{r}, t)\Psi(\mathbf{r}, t)$ to $\Psi^*(\mathbf{r}, t_0)\Psi(\mathbf{r}, t_0)$ and implies a certain kind of selection; the occurrence of the reference time τ is obviously one of the features for that.

How does nature organize that, i.e., the transition from the situation defined by Eq. (F41) to that given by Eq. (F42)? This cannot be due to an internal transformation within the system alone, i.e., without the influence of another process. Beside the system under investigation only the observer's function appears within the frame of our analysis and, therefore, the transition from $\Psi^*(\mathbf{r}, t)\Psi(\mathbf{r}, t)$ to $\Psi^*(\mathbf{r}, t_0)\Psi(\mathbf{r}, t_0)$ must be due to an interplay between the system, described by $\Psi(\mathbf{r}, t)$, and the observer's function. However, the observer's function has been characterized so far by one parameter only: It is the reference time τ, which is measured by our clocks in everyday life. Clearly, only one parameter (τ) is not sufficient for the description of the interplay between the system characterized by $\Psi(\mathbf{r}, t)$ and the observers function. How can we characterize the observation process more realistically? The answer is straightforward.

Let us define a reference system that is produced inside the observer, and let us formally describe it by the wave function $\Psi_{ref}(t)$ and the probability distribution $\Psi^*_{ref}(t)\Psi_{ref}(t)$, respectively, and we would like to characterize the time variable for the reference system by $\gamma = t$. Clearly, $\gamma = t$ is a system-specific quantity. (For simplicity we would like to assume that the reference system is not dependent on any position $\mathbf{r}$.)

The source for the existence of the wave function $\Psi_{ref}(t)$ are energy fluctuations in the brain's fictitious reality [$(\mathbf{p}, E)$-space] leading to $\Psi_{ref}(E)$ and $\Psi^*_{ref}(E)\Psi_{ref}(E)$, respectively. $\Psi_{ref}(E)$ and $\Psi_{ref}(t)$ are connected by [1]

$$\Psi_{ref}(E) = \frac{1}{(2\pi\hbar)^{1/2}} \int_{-\infty}^{\infty} \Psi_{ref}(t) \exp\left\{i\frac{E}{\hbar}t\right\} dt. \tag{F43}$$

Essentially, at time τ we have two probability distributions, $\Psi^*_{ref}(t)\Psi_{ref}(t)$ and $\Psi^*(\mathbf{r}, t)\Psi(\mathbf{r}, t)$; one for the description of time $\gamma = t$ of the reference system and the other for the description of time t of the system under investigation.

The reference system, characterized by $\Psi_{ref}(t)$, is located inside the brain of the observer. The system under investigation is positioned outside the observer and is characterized within the observer's space and time by $\Psi(\mathbf{r}, t)$. For the recognition (observation) of the system $\Psi(\mathbf{r}, t)$ by $\Psi_{ref}(t)$ both systems must be coupled.

The reference system, described by $\Psi_{ref}(t)$, has two functions:

1. To describe the nature of the reference time more specifically, and
2. To select $\Psi^*(\mathbf{r}, t_0)\Psi(\mathbf{r}, t_0)$ from $\Psi^*(\mathbf{r}, t)\Psi(\mathbf{r}, t)$ (see Eqs. (F41) and (F42)).

Both functions are interconnected: Selection is not possible without the existence of a systematically varying reference system.

So far, we have stated that time τ runs monotonically from the past to the future. However, this time-feeling must also be due to a process (inside the brain of the observer) and is therefore also a system-specific time.

Once again, the reference time is described by the distribution $\Psi^*_{ref}(t)\Psi_{ref}(t)$: The probability of finding a certain value $\gamma = t$ for the reference time in the interval $\Delta\gamma = \Delta t$ around $\gamma = t$ is given by $\Psi^*_{ref}(t)\Psi_{ref}(t)\Delta t$. In other words, the reference time $\gamma = t$ becomes uncertain because $\Psi^*_{ref}(t)\Psi_{ref}(t)$ has a certain width Δ_τ and, therefore, no longer runs strictly from the past to the future as is suggested by our clocks used in everyday life. However, the probability distribution $\Psi^*_{ref}(t)\Psi_{ref}(t)$ for the reference time $\gamma = t$ should be a relatively sharp function and, furthermore, because the time τ of our clocks runs monotonically from the past to the future, the distribution $\Psi^*_{ref}(t)\Psi_{ref}(t)$ must run monotonically from the past to the future and we have

$$\Psi^*_{ref}(t)\Psi_{ref}(t) \rightarrow \Psi^*_{ref}(\tau - t)\Psi_{ref}(\tau - t). \tag{F44}$$

We do not measure the time τ but the variable $\gamma = t$ of the distribution for the reference time, and this $\gamma = t$ is uncertain due to $\Psi^*_{ref}(\tau - t)\Psi_{ref}(\tau - t)$. However, because $\Psi^*_{ref}(\tau - t)\Psi_{ref}(\tau - t)$ can be assumed to be a relatively sharp function, the variable $\gamma = t$ for the distribution of the reference time should be close to τ. Due to τ the whole curve $\Psi^*_{ref}(\tau - t)\Psi_{ref}(\tau - t)$ moves strictly from the past to the future, but the values $\gamma = t$ for the reference time fluctuate around τ.

We have outlined that the existence of the wave function $\Psi_{ref}(t)$ for the reference system should be responsible for selection processes. In fact, the transition from $\Psi^*(\mathbf{r}, t)\Psi(\mathbf{r}, t)$ to $\Psi^*(\mathbf{r}, t_0)\Psi(\mathbf{r}, t_0)$ can be explained on the basis of this kind of function, i.e., by $\Psi_{ref}(t)$. In other words, the interplay between the two systems (the reference system described by $[\Psi_{ref}(t), \Psi_{ref}(E)]$ and, on the other hand, the system under investigation described by $[\Psi(\mathbf{r}, t), \Psi(\mathbf{r}, E)]$ should lead to the selection process. This process obviously filters the configuration $\Psi^*(\mathbf{r}, t_0)\Psi(\mathbf{r}, t_0)$ from $\Psi^*(\mathbf{r}, t)\Psi(\mathbf{r}, t)$ out.

However, the following has to be considered. The reference system, characterized by $\Psi_{ref}(t)$ and $\Psi^*_{ref}(\tau - t)\Psi_{ref}(\tau - t)$, respectively, is located inside the brain of the observer. The system under investigation is positioned outside the observer and is characterized within the observer's space and time by $\Psi(\mathbf{r}, t)$ and $\Psi^*(\mathbf{r}, t)\Psi(\mathbf{r}, t)$, respectively. For the observation of the system $\Psi(\mathbf{r}, t)$ by $\Psi_{ref}(t)$ both systems must be coupled. In [1] realistic models have been proposed, and we came to the following result:

The states $\Psi^*(\mathbf{r}, t)\Psi(\mathbf{r}, t)$ of the system under investigation will be systematically scanned by $\Psi^*_{ref}(\tau - t)\Psi_{ref}(\tau - t)$ and only those values of t which correspond with the reference time τ (see also Fig. F5) can be observed. This leads to an effect of motion. The sense of time τ is to select a certain configuration $\Psi^*(\mathbf{r}, t_k)\Psi(\mathbf{r}, t_k)$ with $t_k = \tau$. Clearly, $\Psi^*(\mathbf{r}, t)\Psi(\mathbf{r}, t)$ is a static function if it behaves stationary (see Sec. F.4, Fig. F1) and does not change in the course of time τ, and the effect of motion we experience in connection with $\Psi^*(\mathbf{r}, t)\Psi(\mathbf{r}, t)$ is entirely due to the "motion" of the reference time τ.

F.10 TIME AND PARTICLES WITHIN THE USUAL QUANTUM THEORY

F.10.1 Schrödinger's Equations

The system-specific time t is not known in the conventional quantum theory; only the time τ exists, which is measured with our clocks and has nothing to do with the system under investigation. Schrödinger's equations describe the basic quantum states and belong to the fundamentals of the conventional quantum theory. In the case of stationary systems Schrödinger's equation takes the form

$$i\hbar\frac{\partial}{\partial\tau}\psi(\mathbf{r}, \tau) = -\frac{\hbar^2}{2m_0}\Delta\psi(\mathbf{r}, \tau) + U(x, y, z)\psi(\mathbf{r}, \tau), \tag{F45}$$

with

$$U(x, y, z) \to U(x, y, z, \tau) \tag{F46}$$

we obtain the non-stationary Schrödinger's equation:

$$i\hbar\frac{\partial}{\partial\tau}\psi(\mathbf{r}, \tau) = -\frac{\hbar^2}{2m_0}\Delta\psi(\mathbf{r}, \tau) + U(x, y, z, \tau)\psi(\mathbf{r}, \tau), \tag{F47}$$

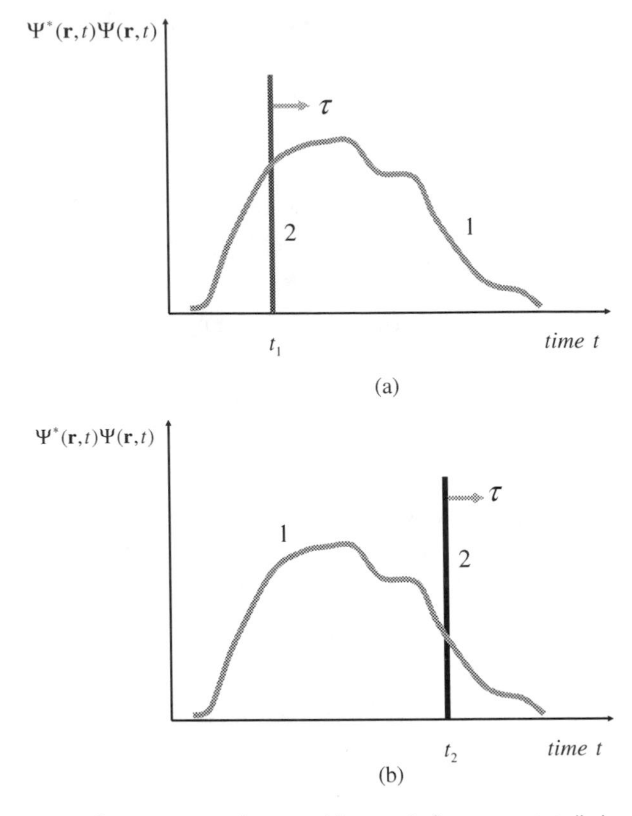

Fig. F5 The system under investigation, characterized by curve 1, fluctuates statistically between all possible configurations t [defined by the range Δ_t of $\Psi^*(\mathbf{r}, t)\Psi(\mathbf{r}, t)$]. On the other hand, curve 2, described by $\Psi^*_{ref}(\tau - t)\Psi_{ref}(\tau - t)$, characterizes the observers time feeling and this must also be due to a process. Δ_τ is the width of $\Psi^*_{ref}(\tau - t)\Psi_{ref}(\tau - t)$ and must be relatively small. The states $\Psi^*(\mathbf{r}, t)\Psi(\mathbf{r}, t)$ of the system under investigation are systematically scanned by $\Psi^*_{ref}(\tau - t)\Psi_{ref}(\tau - t)$ and only those configurations of t, which correspond with the reference time τ, are selected. This leads to an effect of motion. The sense of time τ is to select a certain configuration $\Psi^*(\mathbf{r}, t_k)\Psi(\mathbf{r}, t_k)$ with $t_k = \tau$. [Fig. (a): $\tau = t_1$, Fig. (b):$\tau = t_2$] Clearly, $\Psi^*(\mathbf{r}, t)\Psi(\mathbf{r}, t)$ is a static function if it behaves stationary (see Sec. F.4, Fig. F1) and does not change in the course of time τ, and the effect of motion we experience in connection with $\Psi^*(\mathbf{r}, t)\Psi(\mathbf{r}, t)$ is entirely due to the "motion" of the reference time.

where $U(x, y, z, \tau)$ is the classical interaction potential. Schrödinger's equations describe quantum systems within the frame of the container principle. Here we have exclusively "distance-dependent" interactions. In contrast to the projection theory, form interactions for the description of the shape (form) of the interacting particles are not defined here. Thus, the elements $\mathbf{r} = (x, y, z)$ in Eqs. (F45) and (F46) reflect relative distances between the particles, i.e., we have for example $x \rightarrow x_2 - x_1$.

Once again, the time τ appears within Schrödinger's theory as an external parameter and time t is not defined here. Time is still a classical quantity in the conventional quantum theory. But we have uncertainties $\delta p_x, \ldots$ and $\delta x, \ldots$ forming the well-known uncertainty relations

$$\delta p_x \delta x \geq \frac{\hbar}{2}, \; \delta p_y \delta y \geq \frac{\hbar}{2}, \; \delta p_z \delta z \geq \frac{\hbar}{2}, \qquad \text{(F48)}$$

which are identical with the result of the projection theory (Eq. (F11)). However, the corresponding equation for the time and the energy $\delta E \delta t \geq \hbar/2$ (Eq. (F12)) is not defined in the conventional quantum theory.

In other words, there is no uncertainty relation for the energy and time which would agree in its physical content with the uncertainty relation for the coordinates and momenta; we have the uncertainties $\delta p_x, \ldots$ and $\delta x, \ldots$, but there is no uncertainties with respect to the energy E and the time τ. The significance of the well-known relation

$$\Delta E \Delta \tau \geq \frac{\hbar}{2} \qquad \text{(F49)}$$

(see also Appendix C) is entirely different from that of Eq. (F48). This difference is symbolically expressed by the use of Δ instead of δ. The energy E can be measured in the usual quantum theory to any degree of accuracy at any instant. The quantity ΔE in Eq. (F49) is the difference between two exactly measured values of the energy at two different instants and is not the uncertainty in the value of the energy at a given instant (see also the discussion in [40] and [41]). The same is true for the classical time τ: There is definitely no uncertainty with respect to τ. In a nutshell, in contrast to the projection theory, there is no quantum-aspect of time in the traditional quantum theory.

In Mario Bunge's opinion the energy-time relation (F49) *is a total stranger to quantum theory* [42]. In particular, we find in [42]: *This relation is made plausible by reference to some thought experiments,* *to radioactive decay, and to line breadths. But unlike the genuine indeterminacy relations, $\Delta E \Delta \tau \geq \hbar/2$ has never been proved from first principles. In other words, $\Delta E \Delta \tau \geq \hbar/2$ does not belong to quantum theory, but is just a piece of doubtful heuristics.*

F.10.2 Individuals

Not only in connection with time τ that the conventional quantum theory behaves classically, but, in a certain sense, classical elements obviously come also into play when we analyze the basic quantum events in "space". Within the conventional quantum theory matter is embedded in space, i.e., the container principle is assumed to be valid (Chap. 1) and, within Born's probability interpretation of quantum theory, we need a particle of a certain mass, say m_0, which is embedded in space.

What does the momentum-position uncertainty relation (F48) mean in connection with a trajectory of a particle? The answer is straightforward: Because the quantities $\delta p_x, \ldots$ and $\delta x, \ldots$ reflect the uncertainties in the values of the momenta and the coordinates at the same instant, (F48) implies that there cannot exist a "trajectory" for a particle in space. Since the initial values of x, y, z and p_x, p_y, p_z are inherently uncertain, so is the future trajectory of the particle underdetermined; it is not defined. This has two essential consequences:

1. The existence of a particle (local existent) without a trajectory means that, within the frame of the conventional quantum theory, the motion must be accidental leading to the probability interpretation proposed by Born.
2. One of the consequences of this picture is that there can be no such concept as the *velocity* of a particle in the classical sense of the word, i.e., the limit to which the difference of the two coordinates at two instants, divided by the interval $\Delta\tau$ between these instants, tends to infinity as $\Delta\tau$ tends to zero.

Born proposed to interpret the Schrödinger's wave function $\psi = \psi(x, y, z, \tau)$ not as a true material wave, but as a probability wave.

According to him, the probability of finding a real quantum object (assumed to be a point-like particle of mass m_0) in the volume element dV at a certain position x, y, z and time τ is given by

$$\psi^*(x, y, z, \tau)\psi(x, y, z, \tau)\, dV. \quad \text{(F50)}$$

Thus, $\psi^*\psi$ plays the role of a probability density. It must however be emphasized that $\psi^*\psi$ can only be defined in connection with a real particle of mass m_0, which is embedded in space.

Born's probability interpretation is one of the fundamentals of the usual quantum theory. However, within relativistic quantum mechanics the wave function cannot be used for the definition of a probability density for a single particle as in the case of the (non-relativistic) Schrödinger equation.

In summary, we have a particle without trajectory, and the classical concept of velocity cannot be used here. Therefore, the particle motion must be accidental leading to the probability interpretation for the wave function $\psi(x, y, z, \tau)$. In other words, there is no longer a physical law that tells us *when* and *where* a particle jumps. Within classical mechanics motion is a *continuous blend of changing positions. The object moves in a flow from one point to another. Quantum mechanics failed to reinforce that picture. In fact, it indicated that motion could not take place in that way. They jumped from one place to another, seemingly without effort and without bothering to go between two places* [43]. This text suggests that we should hesitate to connect such a behaviour with a particle in the conventional (classical) sense of the word. It is possibly better to talk only of an "event" without imaging the details. However, it is not conceivable how the classical particle concept can be given up within the usual quantum theory. On the other hand, we have to ask whether it gives sense at all to assume that in nature there exist a localized particle without trajectories.

F.10.3 Particle without Inertia

These arguments become more accessible when we consider the particle-problem in terms of typical observations in connection with the probability density $\psi^*(x, y, z, \tau)\psi(x, y, z, \tau)$. Let

$$\psi^*(x, y, z, \tau)\psi(x, y, z, \tau) \neq 0 \tag{F51}$$

be in the range Δx, and let A and B two positions within Δx. At time τ_A we register the particle with a detector at position A; at time τ_B we register the same particle at position B. Because there is no physical law that tells us "when" the real particle with mass m_0 arrives at position B after it has left position A, the time interval $\Delta\tau = \tau_B - \tau_A$ can take any value, for example, it may be infinitesimal small but different from zero. Then, the apparent velocity $\upsilon = \Delta x/\Delta\tau$ can take any value if Δx is not infinitesimal small, but it can be as small as the diameter of the hydrogen atom. Thus, the velocity $\upsilon = \Delta x/\Delta\tau$ can be close to infinity and may be larger than the velocity of light c and we may write

$$\upsilon \Rightarrow \infty \tag{F52}$$

This peculiarity means that this quantum-mechanical particle of mass m_0 cannot have the property of inertia. Although the particle is embedded in space, the space has no influence on it.

As we have pointed out in Chap. 1, within Newton's mechanics the space is the source of inertia, but this is an ill construction. As we remarked several times, the space cannot create physically real effects because its elements are not observable, and the effect of inertia reflects a physically real effect.

Since the quantum-mechanical particle of mass m_0 does not show the effect of inertia, the space need not give rise to a physically real effect (inertia) and this alone is required when we consider the basic features of space (see in particular Chap. 1). In other words, in the transition from classical mechanics to the conventional quantum theory the ill construction of a space-induced inertia is obviously eliminated. Mach's principle is obviously fulfilled within Born's probability interpretation of quantum phenomena. As we have already pointed out in Chap. 1, Mach required that the space can never be the source for physically real effects, i.e., the space can never act on material objects giving them certain properties as, for example, inertia. According to Mach, a particle does not move in un-accelerated motion relative to space, but relative to the centre of all the other masses in the universe. The particle has to interact with other masses in the universe in order to get the peculiarity of inertia.

This means that the transition from classical mechanics to the conventional quantum theory must lead to the effect that this inertia-producing interaction of the particle (having the mass m_0) with the other masses in the universe must be switched off simultaneously, and this is because the effect of inertia is switched off. Such a scenario is however hardly possible. In other words, an inertia-producing interaction in the sense of Mach should not exist, as long we work on the basis of the container principle.

All these problems indicate that the interpretation of $\psi^*(x, y, z, \tau)$ $\psi(x, y, z, \tau)$ in connection with a real material mass, which is embedded in space, seems to be an ill construction. This is the case for the conventional quantum theory, but the problems disappear when we enter the projection theory where no real material body is embedded in space and time.

F.11 MOTION OF BODIES IN SPACE AND TIME

Let us consider two systems i and j that are positioned in $(\mathbf{r}, t)$-space, and let us apply the projection theory. The wave functions are given by $\Psi(\mathbf{r}, t)_k$, $k = i, j$. With Eq. (F4) we get for the information in $(\mathbf{p}, E)$-space

$$\Psi(\mathbf{p}, E)_k = \frac{1}{(2\pi\hbar)^2} \int_{-\infty}^{\infty} \Psi(\mathbf{r}, t)_k \times \exp\left\{-i\left[\frac{\mathbf{p}}{\hbar} \cdot \mathbf{r} - \frac{E}{\hbar} t\right]\right\} dxdydzdt, \quad k = i, j. \tag{F53}$$

Let us assume that the systems i and j do not interact via external interactions. The existence of i and j are guaranteed if there are existence-inducing interactions (see Sec. F.6.2) and/or if each system contains certain subsystems; in this case Eq. (F21) has to be considered. Such conditions are necessary; otherwise the systems i and j are not able to exist (see Sec. F.7.1).

We may shift the functions $\Psi(\mathbf{r}, t)_k$ relative to $(\mathbf{r}, t)$-space without having to change the interaction [1], i.e., the laws for the $\mathbf{p}, E$-fluctuations, which are described by the probability densities $\Psi^*(\mathbf{p}, E)_k \Psi(\mathbf{p}, E)_k$, $k = i, j$, are not changed when the variables $\mathbf{r}$ and t are shifted by the quantities $\Delta_{k,\mathbf{r}}$ and $\Delta_{k,t}$:

$$\begin{aligned} \mathbf{r} &\rightarrow \mathbf{r} - \Delta_{k,\mathbf{r}}, \quad k = i, j \\ t &\rightarrow t - \Delta_{k,t}, \quad k = i, j \end{aligned} \tag{F54}$$

where the values $\Delta_{k,\mathbf{r}}$ and $\Delta_{k,t}$ are assumed not to be dependent on $\mathbf{r}$ and t but on the reference time τ:

$$\begin{aligned} \Delta_{k,\mathbf{r}} &= \Delta_{k,\mathbf{r}}(\tau) \\ \Delta_{k,t} &= \Delta_{k,t}(\tau). \end{aligned} \tag{F55}$$

The elements $\Delta_{k,\mathbf{r}}$ and $\Delta_{k,t}$ describe certain shifts of the wave function $\Psi(\mathbf{r}, t)_k$ relative to $(\mathbf{r}, t)$-space.

F.11.1 The p, E-Distributions

We would like to mark the moving systems i and j by S' and S''. The following is shown in [1]: The observers, resting relative to the moving systems S' and S'', measure exactly the same $\mathbf{p}, E$-distribution

$$\Psi^*(\mathbf{p}, E)_k \Psi(\mathbf{p}, E)_k, k = i, j \tag{F56}$$

as the observer, who is resting relative to $(\mathbf{r}, t)$-space. This scenario can be summarized schematically as follows:

$$\Psi^*(\mathbf{p}, E)_k \Psi(\mathbf{p}, E)_k \rightarrow \Psi^*(\mathbf{r} - \Delta_{k,\mathbf{r}}, t - \Delta_{k,t})_k \Psi(\mathbf{r} - \Delta_{k,\mathbf{r}}, t - \Delta_{k,t})_k \tag{F57}$$

for $k = i, j$. In other words, the process in $(\mathbf{p}, E)$-space is not dependent on the space-time positions $\Delta_{k,\mathbf{r}}(\tau)$ and $\Delta_{k,t}(\tau)$ of the geometrical object defined by the probability density

$$\Psi^*(\mathbf{r} - \Delta_{k,\mathbf{r}}, t - \Delta_{k,t})_k \, \Psi(\mathbf{r} - \Delta_{k,\mathbf{r}}, t - \Delta_{k,t}), \; k = i, j, \tag{F58}$$

in $(\mathbf{r}, t)$-space (see in particular [1, 2]). In other words, the motion, characterized by $\Delta_{k,\mathbf{r}}$ and $\Delta_{k,t}$, can be arbitrary and has no influence on the real properties of the systems. In particular, the functions

$\Psi(\mathbf{r}, t)_k$, $k = i, j$, are shifted by the quantities $\Delta_{k,\mathbf{r}}$ and $\Delta_{k,t}$ but its form (shape) remain conserved. Then, we may conclude that it is a "motion relative to nothing" [1]. What does the term "motion relative to nothing" mean?

Since the systems i and j do not interact with external entities, the parameters $\Delta_{k,\mathbf{r}}$ and $\Delta_{k,t}$, $k = i, j$, behave like the variables of free systems. In this case the wave functions $\Psi(\Delta_{k,\mathbf{r}}, \Delta_{k,t})_k$, $k = i, j$, are given by [1]

$$\Psi(\Delta_{k,\mathbf{r}}, \Delta_{k,t})_k = C\,\Psi(\mathbf{p}_{k,\mathbf{r}}, E_{k,t})_k \exp\left\{\frac{i}{\hbar}\left[\mathbf{p}_{k,\mathbf{r}} \cdot \Delta_{k,\mathbf{r}} - E_{k,t}\,\Delta_{k,t}\right]\right\}, \tag{F59}$$

where C is a constant and the variables $\mathbf{p}_{k,\mathbf{r}}$, $E_{k,t}$ are the Fourier variables with respect to $\Delta_{k,\mathbf{r}}$ and $\Delta_{k,t}$; $\mathbf{p}_{k,\mathbf{r}}$, $E_{k,t}$ belong to $(\mathbf{p}, E)$-space and $\Delta_{k,\mathbf{r}}$, $\Delta_{k,t}$ to $(\mathbf{r}, t)$-space; a more detailed discussion is given in [1].

The values of $\Delta_{k,\mathbf{r}} = \Delta_{k,\mathbf{r}}(\tau)$ and $\Delta_{k,t} = \Delta_{k,t}(\tau)$ in $(\mathbf{r}, t)$-space are determined by the probability densities $\Psi^*(\Delta_{k,\mathbf{r}}, \Delta_{k,t})_k \Psi(\Delta_{k,\mathbf{r}}, \Delta_{k,t})_k$, which are given in the case of (F59) by the expressions

$$\Psi^*(\Delta_{k,\mathbf{r}}, \Delta_{k,t})_k \Psi(\Delta_{k,\mathbf{r}}, \Delta_{k,t})_k = C^2\,\Psi^*(\mathbf{p}_{k,\mathbf{r}}, E_{k,t})_k \Psi(\mathbf{p}_{k,\mathbf{r}}, E_{k,t})_k, \tag{F60}$$

i.e., the probability densities is independent of $\Delta_{k,\mathbf{r}}$ and $\Delta_{k,t}$ — while $\Delta_{k,\mathbf{r}}$ and $\Delta_{k,t}$ occupy space and time uniformly; no space-time position is preferred. The elements $\Delta_{k,\mathbf{r}}(\tau)$ and $\Delta_{k,t}(\tau)$ fluctuate arbitrarily as a function of time τ.

To sum up, the systems i and j, characterized in $(\mathbf{r}, t)$-space by

$$\Psi^*(\mathbf{r} - \Delta_{k,\mathbf{r}}, t - \Delta_{k,t})_k\,\Psi(\mathbf{r} - \Delta_{k,\mathbf{r}}, t - \Delta_{k,t}), \quad k = i, j$$

are projected onto $(\mathbf{r}, t)$-space arbitrarily, and there are no systematic jumps with respect to time τ. These moving systems i and j, which we mark by S' and S'', move arbitrarily together with the observers, resting relative to S' and S''.

What can we say about an observer, who is resting relative to a frame of reference S, identical with $(\mathbf{r}, t)$-space? What can he say about the jumping systems i and j from the point of view of S? If we allow that the position $\Delta_{k,\mathbf{r}}(\tau)$ and the time $\Delta_{k,t}(\tau)$ take all possible values ($-\infty \leq \Delta_{k,\mathbf{r}}, \Delta_{k,t} \leq \infty$), the functions $\Psi(\Delta_{k,\mathbf{r}}, \Delta_{k,t})_k$ and the probability densities must be zero [1]:

$$\Psi(\Delta_{k,\mathbf{r}}, \Delta_{k,t})_k = 0$$

$$\Psi^*(\Delta_{k,\mathbf{r}}, \Delta_{k,t})_k \Psi(\Delta_{k,\mathbf{r}}, \Delta_{k,t})_k = 0 \qquad \text{(F61)}$$

with $k = i, j$. More details concerning the treatment of free, non-interacting systems are given in [1].

An observer, who is resting in the frame of reference S, is not able to observe the systems i and j; only the observers in the moving frames S' and S'' can give experimental statements about the systems i and j.

From all these features we have to conclude that the quantities $\Delta_{k,\mathbf{r}}$ and $\Delta_{k,t}$ must behave statistically; both functions $\Delta_{k,\mathbf{r}}$ and $\Delta_{k,t}$ do not vary systematically with time τ, and this is because such a law does not exist. If we have certain values $\Delta_{k,\mathbf{r}}(\tau_1)$ and $\Delta_{k,t}(\tau_1)$ at time τ_1, we cannot say something about the values $\Delta_{k,\mathbf{r}}$ and $\Delta_{k,t}$ at time τ_2, even when τ_2 is the next time-value after τ_1, i.e., the quantities $\Delta_{k,\mathbf{r}}(\tau_2)$ and $\Delta_{k,t}(\tau_2)$ can principally not be predicted.

In summary, the quantities $\Delta_{k,\mathbf{r}}$ and $\Delta_{k,t}$ jump statistically through $(\mathbf{r}, t)$-space together with the space-time positions of the probability densities, defined by

$$\Psi^*(\mathbf{r} - \Delta_{k,\mathbf{r}}(\tau), t - \Delta_{k,t}(\tau))\Psi(\mathbf{r} - \Delta_{k,\mathbf{r}}(\tau), t - \Delta_{k,t}(\tau)) \,, k = i, j, \qquad \text{(F62)}$$

i.e., both structures jump arbitrarily through space and time. These jumps are independent from each other.

The projection of $\Psi(\mathbf{p}, E)_k$, $k = i, j$, onto $(\mathbf{r}, t)$-space leads to the wave functions

$$\Psi(\mathbf{r} - \Delta_{k,\mathbf{r}}, t - \Delta_{k,t}), \quad k = i, j,$$

and the geometrical structures (probability densities), given by Eq. (F62), should be considered as a definition of the forms (shapes) of the systems i and j; and, as we have outlined above, these geometrical structures jump arbitrarily relative to $(\mathbf{r}, t)$-space. In the following let us briefly repeat and deepen the physical content of this effect.

F.11.2 Effective Velocities

We want to assume that system i is at time τ_a at the space-time position $\mathbf{r}_{1a}, t_{1a}$ and that system j is at the same time, τ_a, at the space-time position $\mathbf{r}_{2a}, t_{2a}$. Furthermore, let us assume that system i is at time $\tau_b \neq \tau_a$ at the space-time position $\mathbf{r}_{1b}, t_{1b}$ and system j is at the same time, τ_b, at $\mathbf{r}_{2b}, t_{2b}$. Then, the following is relevant: There is no law defined that would predict the values $\mathbf{r}_{1b}, t_{1b}$ on the basis of $\mathbf{r}_{1a}, t_{1a}$ and there is of course also no law that would predict the values $\mathbf{r}_{2b}, t_{2b}$ on the basis of $\mathbf{r}_{2a}, t_{2a}$. In other words, both systems take arbitrary space-time positions in the course of time τ.

Because there is no distance-dependent correlation effective between system i and system j, both structures (peaks) jump independently from each other through $(\mathbf{r}, t)$-space, i.e., there can be no relationship between the space-time positions $\mathbf{r}_{1a}, t_{1a}$ and $\mathbf{r}_{2a}, t_{2a}$ and also none between $\mathbf{r}_{1b}, t_{1b}$ and $\mathbf{r}_{2b}, t_{2b}$. In other words, there are no relationships (physical laws) between the space-time positions $\mathbf{r}_{1a}, t_{1a}$, $\mathbf{r}_{1b}, t_{1b}$, $\mathbf{r}_{2a}, t_{2a}$ and $\mathbf{r}_{2b}, t_{2b}$, even when τ_b is the next time-value after τ_a. Therefore, the values $\mathbf{r}_{2b}, t_{2b}$ and $\mathbf{r}_{2a}, t_{2a}$ can principally not be predicted. That is, all the quantities $\mathbf{r}_{1a}, t_{1a}$, $\mathbf{r}_{1b}, t_{1b}$, $\mathbf{r}_{2a}, t_{2a}$ and $\mathbf{r}_{2b}, t_{2b}$ are completely independent from each other and behave strictly statistically.

However, we have to keep in mind that within projection theory the peaks (geometrical structures), defined by Eq. (F62), do not move through space-time [$(\mathbf{r}, t)$-space] but are projected on it. Nevertheless, we may define "effective velocities" which we would like to call $v_{m\mathbf{r}}$ and v_{mt}; $v_{m\mathbf{r}}$ is the effective velocity with respect to the variable $\mathbf{r}$ and is given by

$$v_{m\mathbf{r}} = \frac{\mathbf{r}_{mb} - \mathbf{r}_{ma}}{\tau_b - \tau_a}, \quad m = 1, 2 \tag{F63}$$

v_{mt} is the effective velocity with respect to the variable t and has the form

$$v_{mt} = \frac{t_{mb} - t_{ma}}{\tau_b - \tau_a}, \quad m = 1, 2 \tag{F64}$$

The time interval $\tau_b - \tau_a$ can be close to zero and, on the other hand, $\mathbf{r}_{mb} - \mathbf{r}_{ma}$ and $t_{mb} - t_{ma}$ may be as large as the universe, where "large" really means with respect to the maximum space-extension as well as with respect to the maximum time-extension (that is, from the beginning to the end of time). With

$$\tau_b - \tau_a \to 0 \tag{F65}$$

we obtain for the velocities

$$\begin{aligned} v_{m\mathbf{r}} &\to \infty, \quad m = 1, 2 \\ v_{mt} &\to \infty, \quad m = 1, 2. \end{aligned} \tag{F66}$$

In other words, it is a motion without inertia. The situation is illustrated in Fig. F6.

In general, we may express the arbitrary behaviour of the $(\mathbf{r}, t)$-structures as follows: In Eq. (F55) the quantities $\Delta_{k,\mathbf{r}}$ and $\Delta_{k,t}$ are given as a function of time τ. If $\Delta_{k,\mathbf{r}}$ and $\Delta_{k,t}$ take at time τ' the values $\Delta_{k,\mathbf{r}}(\tau')$ and $\Delta_{k,t}(\tau')$ and at time τ'' the values $\Delta_{k,\mathbf{r}}(\tau'')$ and $\Delta_{k,t}(\tau'')$, the quantities

$$\Delta_{k,t}\, r(\tau'') - \Delta_{k,t}\, r(\tau'), \quad k = i, j \tag{F67}$$

and

$$\Delta_{k,t}(\tau'') - \Delta_{k,t}(\tau'), \quad k = i, j \tag{F68}$$

may be expressed by

$$\Delta_{k,\mathbf{r}}(\tau'') - \Delta_{k,\mathbf{r}}(\tau') = v_{k\mathbf{r}}(\tau'' - \tau') + b_{k\mathbf{r}}(\tau'' - \tau')^2/2 + \cdots, \quad k = i, j \tag{F69}$$

$$\Delta_{k,t}(\tau'') - \Delta_{k,t}(\tau') = v_{kt}(\tau'' - \tau') + b_{kt}(\tau'' - \tau')^2/2 + \cdots, \quad k = i, j \tag{F70}$$

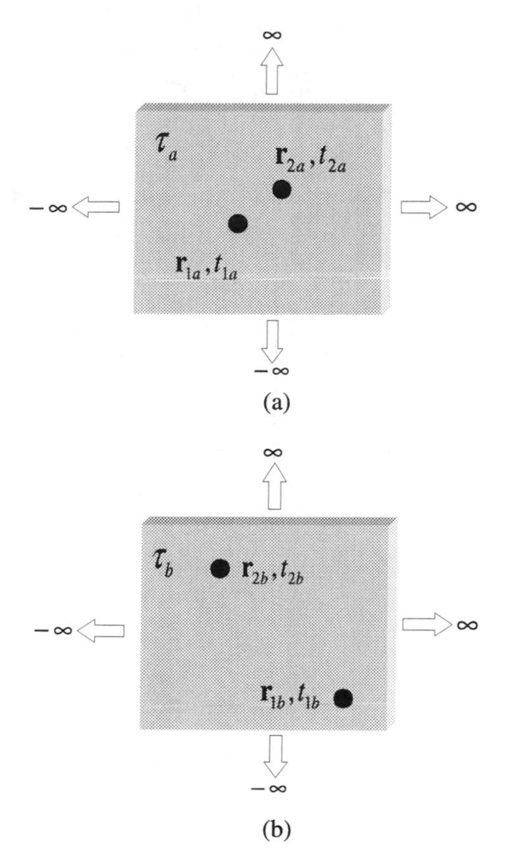

Fig. F6 (a) At time τ_a system i is at the space-time position $\mathbf{r}_{1a}$, t_{1a} and system j at $\mathbf{r}_{2a}$, t_{2a}. (b) At time τ_b system i is at the space-time position $\mathbf{r}_{1b}$, t_{1b} and system j at $\mathbf{r}_{2b}$, t_{2b}. Because of $\tau_b - \tau_a \rightarrow 0$ and the fact that the distances $\mathbf{r}_{mb} - \mathbf{r}_{ma}$ and $t_{mb} - t_{ma}$ (with $m = 1, 2$) may be arbitrarily large (in principle, as large as the space-time extension of the universe), the velocities $\upsilon_{m\mathbf{r}}$ and υ_{mt} may be close to infinity.

where $\upsilon_{m\mathbf{r}}$ and υ_{mt} are again velocities and $b_{k\mathbf{r}}$ and b_{kt} are accelerations. With $\tau'' - \tau' \rightarrow 0$ and arbitrary values for $\Delta_{k,\mathbf{r}}(\tau'') - \Delta_{k,\mathbf{r}}(\tau')$ and $\Delta_{k,t}(\tau'') - \Delta_{k,t}(\tau')$, the values for $\upsilon_{m\mathbf{r}}$ and υ_{mt} and for $b_{k\mathbf{r}}$ and b_{kt} may take arbitrary values but must fulfil the relations (F69) and (F70). In particular, they may be close to infinity:

$$\upsilon_{m\mathbf{r}}, \upsilon_{kt}, b_{kt}, b_{kt} \rightarrow \infty, \quad k = 1, 2 \tag{F71}$$

(Here we have assumed that only the first two terms in equations (F69) and (F70) are different from zero.)

In summary, the arbitrary jumps of the geometrical structures contain velocities $v_{m\mathbf{r}}$ and v_{mt} and accelerations $b_{k\mathbf{r}}$ and b_t; and of course other types of components when we consider more than two terms in the relations (F69) and (F70). As we have mentioned several times, these components have no influence on the $\mathbf{p}$, E-states of both systems i and j [1].

F.11.3 Motion Relative to Nothing

We pointed out above that we may shift the functions $\Psi(\mathbf{r}, t)_k, k = i, j$ (relative to $(\mathbf{r}, t)$-space) without having to change the interaction, which take place in $(\mathbf{p}, E)$-space; i.e., the laws for the $\mathbf{p}, E$-fluctuations, which are described by the probability densities $\Psi^*(\mathbf{p}, E)_k\Psi(\mathbf{p}, E)_k, k = i, j$, are not changed when $\mathbf{r}$ and t are shifted by the quantities $\Delta_{k,\mathbf{r}}$ and $\Delta_{k,t}$ (see Eq. (F54):

$$\mathbf{r} \to \mathbf{r} - \Delta_{k,\mathbf{r}}, k = i, j$$

$$t \to t - \Delta_{k,t}, k = i, j$$

Then, $\Psi(\mathbf{r}, t)_k$ can be reformulated as follows:

$$\Phi(\mathbf{p}, E)_k = \frac{1}{(2\pi\hbar)^2}\int_{-\infty}^{\infty} \Psi(\mathbf{r} - \Delta_{k,\mathbf{r}}, t - \Delta_{k,t}) \times \exp\left\{-i\left[\frac{\mathbf{p}}{\hbar}\cdot\mathbf{r} - \frac{E}{\hbar}t\right]\right\} dxdydzdt, \quad k = i, j \tag{F72}$$

with

$$\Phi(\mathbf{p}, E)_k = \Psi(\mathbf{p}, E)_k \exp\left\{-i\left[\frac{\mathbf{p}}{\hbar}\cdot\Delta_{k,\mathbf{r}} - \frac{E}{\hbar}\Delta_{k,t}\right]\right\}, \quad k = i, j \tag{F73}$$

and we immediately obtain

$$\Phi^*(\mathbf{p}, E)_k\Phi(\mathbf{p}, E)_k = \Psi^*(\mathbf{p}, E)_k\Psi(\mathbf{p}, E)_k, \quad k = i, j \tag{F74}$$

Because of this property both functions, $\Psi(\mathbf{p}, E)_k$ and $\Phi(\mathbf{p}, E)_k$, are "equivalent" with respect to the description of interaction processes; they are equally suitable to describe $\mathbf{p}$, E-fluctuations.

The arbitrary jumps in connection with the quantities $\Delta_{k,\mathbf{r}} = \Delta_{k,\mathbf{r}}(\tau)$ and $\Delta_{k,t} = \Delta_{k,t}(\tau)$ reflect a certain dynamics of the systems i and j relative to $(\mathbf{r}, t)$-space. The existence of the effective velocities $v_{m\mathbf{r}}$ and v_{mt} (defined by the Eqs. (F63) and F64)) and their possible changes in the course of time τ suggest that the $\mathbf{p}$, E-states of both systems, that are seemingly moving relative to $(\mathbf{r}, t)$-space, are changed. However, this is not the case as we have discussed by means of the functions $\Phi^*(\mathbf{p}, E)_k\Phi(\mathbf{p}, E)_k$ and $\Psi^*(\mathbf{p}, E)_k\Psi(\mathbf{p}, E)_k$ with $k = i, j$ (see Eq. (F74)). This property is confirmed when we investigate the mean values for the momentum $\overline{\mathbf{p}}_k$ and the mean energy $\overline{E}_k$. Both quantities, that is, $\overline{\mathbf{p}}_k$ and $\overline{E}_k$, are completely independent on the (arbitrary) motion of the systems i and j in $(\mathbf{r}, t)$-space. In fact, we have

$$\begin{aligned}\overline{\mathbf{p}}_k &= \int_{-\infty}^{\infty} \mathbf{p}\,\Psi^*(\mathbf{p}, E)_k\Psi(\mathbf{p}, E)_k dp_x dp_y dp_z dE \\ &= \int_{-\infty}^{\infty} \mathbf{p}\,\Phi^*(\mathbf{p}, E)_k\Phi(\mathbf{p}, E)_k dp_x dp_y dp_z dE, \quad k = i, j \qquad \text{(F75)}\end{aligned}$$

and

$$\begin{aligned}\overline{E}_k &= \int_{-\infty}^{\infty} E\,\Psi^*(\mathbf{p}, E)_k\Psi(\mathbf{p}, E)_k dp_x dp_y dp_z dE \\ &= \int_{-\infty}^{\infty} E\,\Phi^*(\mathbf{p}, E)_k\Phi(\mathbf{p}, E)_k dp_x dp_y dp_z dE, \quad k = i, j \qquad \text{(F76)}\end{aligned}$$

with the condition

$$\begin{aligned}&\int_{-\infty}^{\infty} \Psi^{\bullet}(\mathbf{p}, E)_k\Psi(\mathbf{p}, E)_k dp_x dp_y dp_z dE \\ &\quad = \int_{-\infty}^{\infty} \Phi^{\bullet}(\mathbf{p}, E)_k\Phi(\mathbf{p}, E)_k dp_x dp_y dp_z dE = 1, \quad k = i, j \qquad \text{(F77)}\end{aligned}$$

In other words, the arbitrary jumps have no influence on the real material properties of the systems. Then, we may conclude that it is a "motion relative to nothing".

The motion of the geometrical structures in $(\mathbf{r}, t)$-space with certain effective velocities $v_{m\mathbf{r}}$ and v_{mt} (Eqs. F(63) and (F64)) does not lead to changes in the $\mathbf{p}$, E-states of the systems i and j. We have characterized this situation by the statement "motion relative to nothing". This is in contrast to the facts in conventional physics. Here a certain velocity, say $\mathbf{v}$, determines the momentum $\mathbf{p}$ and the energy E. We have $\mathbf{p} = m_0\mathbf{v}$ and $E = m_0\mathbf{v}^2/2$.

The physical picture within Newton's mechanics is quite different from that of the projection theory. In classical mechanics the $\mathbf{p}$, E-states are definitely changed when a body moves through space (relative to space) with varying velocities, and this is explained in Newton's mechanics by space-effects. However, such a kind of space-effect has to be considered as unphysical. As we have pointed out in Chap. 1, space and time cannot be the source of physically real effects.

In fact Newton's conception of space has been strongly criticized by Ernst Mach and, in particular, also by Albert Einstein. Nevertheless, within the Special Theory of Relativity and the General Theory of Relativity the situation concerning space-effects could not really improved.

Within Newton's theory the notion of "inertia" is of particular relevance. Here all real bodies are embedded in space. Even when a body does not interact with other bodies, the effect of inertia is effective, i.e., the body moves through space with constant velocity $\mathbf{v}$ where the velocity of zero is included, and the effect of inertia is entirely due to the interaction of the body with space. The body moves relative to space, and its $\mathbf{p}$, E-state is dependent on its velocity relative to space, i.e., the variation of $\mathbf{v}$ leads to changes of the body's $\mathbf{p}$, E-states.

As we have just recognized, within the projection theory the situation is quite different: The velocity of moving geometrical structures (for example, systems i and j) has no influence on the $\mathbf{p}$, E-states, whereby the velocity can even be infinite (Eq. (F66)). There is no inertia within the projection theory at the basic level; more details are given in Chap. 4.

All the features, which we have pointed out with respect to $\Psi(\mathbf{r}, t)$ are also valid for a closed system consisting of N subsystems (particles), which does not interact with any other external systems, i.e., there are only interactions between the N subsystems. Such a transition can be formally expressed by (F21). The whole system with N subsystems moves arbitrarily through space and time, and this behaviour is also described by the variables $\Delta_{k,\mathbf{r}}(\tau)$ and $\Delta_{k,t}(\tau)$ as in the case of $\Psi(\mathbf{r}, t)$.

Our discussion has been based on systems, which can be described by a wave function $\Psi(\mathbf{r}, t)$; such systems interact in $(\mathbf{p}, E)$-space in such a way that the correlations with other systems in $(\mathbf{r}, t)$-space are "distance-independent". No doubt, the possibility for infinite velocities suggests this kind of interaction.

Remark

We know from the Special Theory of Relativity that the velocity of a body having the rest mass m_0 cannot exceed the velocity of light. However, the property $v_{m\mathbf{r}} \to \infty$, $m = 1, 2$ (Eq. (F66)) seems to be in contradiction to this relevant statement of the Special Theory of Relativity. But this is not the case. The reason is obvious: The observer, resting in system i, cannot say anything about the space-time positions of system j and vice versa. Therefore, a relative motion of both systems is not definable. Furthermore, the velocities $v_{m\mathbf{r}}$, $k = i, j$, define the motion of the systems i and j relative to $(\mathbf{r}, t)$-space, but an observer, who is resting relative to $(\mathbf{r}, t)$-space, is not able to observe the systems i and j, and this fact is expressed by Eq. (F61). Therefore, the property $v_{m\mathbf{r}} \to \infty$, $m = 1, 2$ is not observable.

It must be emphasized that the particles do not move through space and time but are projected on them. This is reason why the laws of the Special Theory of Relativity cannot be applied here. In this connection it should also be mentioned that the effective velocity v_{mt} is not defined in the Special Theory of Relativity because the system-specific time t is not defined here.

F.12 INTERACTIONS AND CORRELATIONS

Let us consider a system, which consists of N interacting subsystems. These subsystems may interact in ($\mathbf{p}$, E)-space in such a way that the $\mathbf{r}, t$-correlations between them are "distance-dependent", that is, the mutual correlations between the subsystems are in ($\mathbf{r}$, t)-space dependent on the space-time distances $\mathbf{r}_k - \mathbf{r}_l, t_k - t_l$ (with $k, l = 1, \ldots, N, k \neq l$), where $\mathbf{r}_1, \mathbf{r}_2, \ldots, \mathbf{r}_N, t_1, t_2, \ldots, t_N$ are the positions and system-specific times in ($\mathbf{r}$, t)-space at time τ.

F.12.1 Distance-Independent Interactions

The projection theory opens up the possibility for another kind of interaction in ($\mathbf{p}$, E)-space, leading to correlations in ($\mathbf{r}$, t)-space that are not dependent on space-time distances between the systems and are therefore "distance-independent". In other words, there can be correlations — between two systems, say i and j — where the strength is not dependent on the space-time distances $\mathbf{r}_i - \mathbf{r}_j, t_i - t_j$. Such interactions define the form (shape) of a system. In Sec. F.11 we worked exclusively on the basis of form-interactions.

In conventional physics we also use certain forms for elementary systems: We have point-like particles, strings, branes etc. However, these specific forms had to be assumed in conventional physics and could not be derived. In contrast to these developments, projection theory opens up the possibility to explain (derive) certain elementary forms in nature by means of this new kind of interaction; it leads to distance-independent correlations and create the geometrical form (shape) of systems in ($\mathbf{r}$, t)-space.

In Sec. F.11 we have treated two systems i and j which can interact via existence-inducing $\mathbf{p}, E$-fluctuations in ($\mathbf{p}$, E)-space that have the effect of "distance-independent" correlations in ($\mathbf{r}, t$)-space, and the probability density $\Psi^*(\mathbf{r}, t)\Psi(\mathbf{r}, t)$ defines the form of the systems. Both systems i and j jump arbitrarily in ($\mathbf{r}, t$)-space and their space-time distance at a certain time τ may be as large as the space-extension of the universe, where "large" really means with respect to the maximum space-extension as well as with respect to the maximum time-extension (that is, from the beginning to the end of time). Nevertheless, both systems i and j interact with a constant strength, even when the space-time distance takes the largest possible value. In other words, both systems interact, but this interaction is independent of the actual space-time positions of both systems.

This property reflects the non-local character of the projection theory. Because the interaction processes exclusively take place in $(\mathbf{p}, E)$-space and the structures in $(\mathbf{r}, t)$-space are projections from $(\mathbf{p}, E)$-space onto space-time, a position-independent interaction becomes possible in a quite natural way. Note that within the $(\mathbf{r}, t)$-space no signals are exchanged between the two systems; within the $(\mathbf{r}, t)$-space there are "only" geometrical structures and $\mathbf{r}, t$-correlations, nothing else.

F.12.2 On the Description of Form-Interactions

The form of a system is defined by Eq. (F7)

$$i\hbar\frac{\partial}{\partial t}\Psi(\mathbf{r}, t) = -\frac{\hbar^2}{2m_0}\Delta\Psi(\mathbf{r}, t) + V(x, y, z, t)\Psi(\mathbf{r}, t)$$

The potential function $V(x, y, z, t)$ is used for the description of the geometrical structure $\Psi^*(\mathbf{r}, t)\Psi(\mathbf{r}, t)$ (form of the system) in $(\mathbf{r}, t)$-space, which reflect $\mathbf{r}, t$-correlations and correspond to $\mathbf{p}, E$-fluctuations in $(\mathbf{p}, E)$-space, i.e., the $\mathbf{r}, t$-correlations and $\mathbf{p}, E$-fluctuations take place simultaneously. The fluctuations take place between the system under investigation and its counterpart, otherwise the conservation laws for the momentum and the energy would be violated (Sec. F.6, Sec. F.11, Fig. F3). In other words, the existence of such a counterpart is a necessary condition. These $\mathbf{p}, E$-fluctuations are described by Eq. (F8)

$$E\,\Psi(\mathbf{p}, E) = \frac{\mathbf{p}^2}{2m_0}\Psi(\mathbf{p}, E) + V\left(i\hbar\frac{\partial}{\partial p_x}, i\hbar\frac{\partial}{\partial p_y}, i\hbar\frac{\partial}{\partial p_z}, -i\hbar\frac{\partial}{\partial E}\right)\Psi(\mathbf{p}, E)$$

As we have pointed out above, $\Psi(\mathbf{r}, t)$ and $\Psi(\mathbf{p}, E)$ are connected by a Fourier transform, and both functions contain exactly the same information.

F.12.3 Interaction within Conventional Physics

Here we work within the frame of the container principle. Let us consider here two systems i and j that are embedded as real bodies in space. The variable for the system-specific time t is not defined in conventional physics but only the reference time τ. The real interaction processes exclusively take place in space, and we would like to assume that these two systems interact with each other, but none of them interact with other bodies.

The effect of an interaction between the systems, having at time τ the sharp space positions $\mathbf{r}_i$ and $\mathbf{r}_j$, is that the space positions are changed. In contrast to the projection theory, note that within conventional physics the form (shape) of the two systems i and j have to be assumed and have to be considered as pre-requisite here. There is no conception to deduce the form (shape) of elementary systems in conventional physics.

Within usual physics (Newton's mechanics) there is the following mechanism: The space positions $\mathbf{r}_i$ and $\mathbf{r}_j$ are changed by the interaction process

$$\mathbf{r}_i, \mathbf{r}_j \rightarrow \mathbf{r}_i + \Delta\mathbf{r}_i, \mathbf{r}_j + \Delta\mathbf{r}_j \tag{F78}$$

In the case of point-like systems the bodies take the forms $\delta(\mathbf{r}-\mathbf{r}_i)$ (system i) and $\delta(\mathbf{r}-\mathbf{r}_j)$ (system j) and we have according to the interaction (see Eq. (F78))

$$\textit{system } i : \delta(\mathbf{r} - \{\mathbf{r}_i + \Delta\mathbf{r}_i\}) \tag{F79}$$

$$\textit{system } j : \delta(\mathbf{r} - \{\mathbf{r}_j + \Delta\mathbf{r}_j\}) \tag{F80}$$

It is normally assumed within usual physics that the interaction strength decreases with increasing distance $\mathbf{r}_i - \mathbf{r}_j$ between the systems, and this is intuitively understandable because it is assumed in usual physics that the interaction processes take place in space. Therefore, the assumption that with decreasing potential function the influence of system i on system j and vice versa decreases becomes intuitively understandable.

But we have to be careful. Within the projection theory there cannot be such kind space-time connections since the interaction processes do not take place in $(\mathbf{r}, t)$-space. As we have outlined above, within the projection theory we have "merely" $\mathbf{r}, t$-correlations in $(\mathbf{r}, t)$-space, and the real interaction processes are identified with $\mathbf{p}, E$-fluctuations in $(\mathbf{p}, E)$-space.

F.12.4 Energy and Time Representation

The function $V(x, y, z, t)$, which has the dimension of energy, cannot be assigned to specific positions x, y, z, t in $(\mathbf{r}, t)$-space; here only geometrical structures (space-time positions) are defined, i.e., quantities with a dimension of energy have no place here. There are no real material bodies in $(\mathbf{r}, t)$-space, and the potential $V(x, y, z, t)$ would only make sense in connection with real material bodies. In contrast to the container principle, within the framework of the projection principle energies are exclusively positioned in $(\mathbf{p}, E)$-space.

The purpose of Eq. (F7) is to determine the wave function $\Psi(\mathbf{r}, t)$. As we have pointed out in Chap. 1, Eq. (F7) itself cannot be depicted in $(\mathbf{r}, t)$-space, but it is positioned on another "level of reality" which is higher than that where the geometrical structures are representable (it is level L_2 in Fig. 17). Equation (F7) is a tool for the determination of $\Psi(\mathbf{r}, t)$ within the energy-representation; other representations for the determination of $\Psi(\mathbf{r}, t)$ can be chosen. Let us briefly discuss the so-called time-representation.

When we use Eqs. (F7) and (F8) and formulate the problem within the classical approximation, the uncertainties

$$\delta x, \ldots, \delta t, \delta p_x, \ldots, \delta E \qquad \text{(F81)}$$

in the values $x, \ldots, t, p_x, \ldots, E$ become zero:

$$\delta x, \ldots, \delta t, \delta p_x, \ldots, \delta E \to 0 \qquad \text{(F82)}$$

Then, the variables $x, \ldots, t, p_x, \ldots, E$ behave classically, i.e., without any uncertainty. Since we have in the projection theory the system-specific time t as an additional variable, instead of Eq. (F6), which can be formulated as $E = E(p_x, \ldots, x, \ldots)$, we have

$$E = E(p_x, \ldots, x, \ldots, t) \qquad \text{(F83)}$$

or the equivalent expression

$$t = t(p_x, \ldots, x, \ldots, E). \qquad \text{(F84)}$$

Using the quantum-theoretical rules derived in [1], we obtain after minor manipulations the following equation for the determination of $\Psi(\mathbf{r}, t)$:

$$t\,\Psi(\mathbf{r}, t) = \hat{T}\left(-i\hbar\frac{\partial}{\partial x}, \ldots, x, \ldots, i\hbar\frac{\partial}{\partial t}\right)\Psi(\mathbf{r}, t) \qquad \text{(F85)}$$

where $\hat{T}(-i\hbar\partial/\partial x, \ldots, x, \ldots, i\hbar\partial/\partial t)$ is the time-operator for the system under investigation, which has to be constructed in such a way that Eq. (F85) is compatible to Eq. (F7). Note that the operator

$$\hat{T}\left(-i\hbar\frac{\partial}{\partial x}, \ldots, x, \ldots, i\hbar\frac{\partial}{\partial t}\right) \qquad \text{(F86)}$$

has nothing to do with the operator for the time-coordinate $\hat{t} = -i\hbar\partial/\partial E$ (see (F10), which is independent of the system under investigation).

Equation (F85) is a real alternative to Eq. (F7). While the wave function $\Psi(\mathbf{r}, t)$ is determined by Eq. (F7) within the "energy-representation", it is determined by Eq. (F85) within the "time-representation".

F.12.5 Interaction Potentials are Auxiliary Elements

In the case of form (distance-independent) correlations in $(\mathbf{r}, t)$-space, the "classical" potential picture (Sec. F.12.3) is not applicable. Instead of one distant-dependent potential for both systems, we have in the case of distance-independent correlations "two" potentials, one for the system under investigation, say i, and another for the counterpart j. Thus, instead of Eq. (F7)

$$i\hbar\frac{\partial}{\partial t}\Psi(\mathbf{r}, t) = -\frac{\hbar^2}{2m_0}\Delta\Psi(\mathbf{r}, t) + V(x, y, z, t)\Psi(\mathbf{r}, t)$$

we have

$$i\hbar\frac{\partial}{\partial t}\Psi(\mathbf{r}, t)_k = -\frac{\hbar^2}{2m_0}\Delta\Psi(\mathbf{r}, t)_k + V(x, y, z, t)_k\,\Psi(\mathbf{r}, t)_k, \quad k = i,\ j.$$

This is the reason why the usual interpretation in terms of more or less classical notions cannot be applied in the projection theory. In fact, the potentials takes in $(\mathbf{p}, E)$-space the form of operators

$$V\left(i\hbar\frac{\partial}{\partial p_x}, i\hbar\frac{\partial}{\partial p_y} i\hbar\frac{\partial}{\partial p_z}, -i\hbar\frac{\partial}{\partial E}\right)_k, \quad k = i,\ j$$

(see Eq. (F8)) and this expression is not accessible to visualizable pictures. Therefore, potential functions should be considered as abstract quantities for the determination of the $\mathbf{p}$, E-fluctuation in the $(\mathbf{p}, E)$-space, which appears in the $(\mathbf{r}, t)$-space as correlations between the various space-time positions $\mathbf{r}$, t. All these statements are of course also valid for "distance-dependent" potential functions.

In other words, it is not the potential that is the primary quantity but the $\mathbf{p}$, E-fluctuations and $\mathbf{r}$, t-correlations, respectively. In $(\mathbf{r}, t)$-space we have exclusively geometrical structures, and no real objects are embedded here. Therefore, it is wrong to assume that there is an exchange of information and energy, due to $V(x, y, z, t)_k$, $k = i, j$, between the (geometrical) structures i and j.

This process might differ considerably from those treated within the frame of the usual quantum theory. The reason is simple: Within the projection theory the equation for the determination of $\Psi(\mathbf{r}, t)$ (Eq. (F7)) is more general than Schrödinger's equation (Eq. (F47)) of the usual quantum theory because the function $V(x, y, z, t)$ is in general more complex than the classical potential $U(x, y, z)$ which is used in the usual quantum theory (Sec. F.10). In principle, $V(x, y, z, t)$ could have an imaginary part. The appearance of the function $V(x, y, z, t)$

is a logical consequence of the theoretical structures of the projection theory (see in particular [1]).

Conclusion: Within the projection theory the interaction potential can be at best interpreted as an auxiliary element without any imaginable background. In particular, there are no energy states in $(\mathbf{r}, t)$-space. But within conventional physics the notion "interaction" becomes questionable when we analyze this term in more detail.

F.12.6 Conventional Description of Interactions

Again, within conventional physics the real bodies are embedded in space and there are real energy states leading to a mutual influence between the bodies (between the earth and the sun or between two electrical charges in the case of the hydrogen atom, etc.). Such situations are expressed in the usual quantum theory by Schrödinger's equation (F47), and the potential function $U(x, y, z)$ is in this context in the center.

How does the mutual influence between two bodies come into existence as, for example, between two electrical charges or between the earth and the sun? What mechanism can explain these mutual influences? We have already discussed this point in [1, 2]. In this connection two "pictures" turned out to be of particular relevance: The "proximity effect" and the "action-at-the-distance"; in [1] we noted the following:

It is of principal interest to note that the "proximity effect" and the "action-at-the-distance" are merely "expressions" or interpretations of the gravitational law to $m_1 m_2/r^2$ *and Coulomb's law to* qQ/r^2*. These force laws cannot, however, be derived from these notions. Many people believe that a mechanism, which is composed of many familiar single processes (preferably from everyday life), can explain the mathematical structure of the force laws. What mechanism is, for example, responsible for the fact that the forces expressed by* $m_1 m_2/r^2$ *are inversely proportional to the square of the distance between the masses* m_1 *and* m_2*? As already mentioned, the ideas "proximity effect" and "action-at-the-distance" cannot give the answer to this question,* *since they interpret the relation $m_1 m_2/r^2$, but are not able to explain the mathematical structure of this force law.*

In summary, there is no possibility to explain by a mechanism how the mutual influence between two bodies comes into existence. The notion of "interaction" has therefore to be considered as an irreducible primary property of matter. According to Thomas Kuhn, the fact that there is no possibility to explain the interaction (by a mechanism) has to be considered as problematic because then Newton's gravity takes on an occult quality [1].

There is obviously no possibility to model the interaction processes in conventional physics. The unsuccessful search for an interaction mechanism can however be explained within the projection theory. Potential functions should be considered in the projection theory as abstract quantities for the determination of the $\mathbf{p}, E$-fluctuation in $(\mathbf{p}, E)$-space which appear in $(\mathbf{r}, t)$-space as $\mathbf{r}, t$-correlations between the various space-time positions. The potential function takes the form of an operator in $(\mathbf{p}, E)$-space and this expression is not accessible to visualizable pictures. This is the main reason why potential functions should be considered as abstract quantities.

In other words, it is not the potential that is the primary quantity but the $\mathbf{p}, E$-fluctuations and $\mathbf{r}, t$-correlations, respectively. In $(\mathbf{r}, t)$-space we have exclusively geometrical structures, and no real objects are embedded here. Therefore, it is wrong to assume that there is an exchange of information and energy, due to $V(x, y, z, t)_k$, $k = i, j$, between the (geometrical) structures i and j.

F.12.7 Phenomena in Usual Quantum Theory

We can go a step further and ask for the role of the potential energy $U(x, y, z)$ in the usual quantum theory. Newton's mechanics is a classical theory but its interaction concept is also used in the usual quantum theory; Schrödinger's equation completely contains the notion of potential energy in the classical sense.

It may be doubted whether the classical concept of potential energy can be used in connection with quantum phenomena. In fact, the projection theory clearly shows that this should not be possible. Schrödinger's equation could not be derived, so that an interpretation of its elements, as for example $U(x, y, z)$, is, strictly speaking, not immediately possible. Thus, the whole concept is dependent on the practical success. Schrödinger's equation has been tested unusually well, but the classical explanation of $U(x, y, z)$ can be considered here as problematic. Concerning this point we find a relevant remark in [44]:

"*...While Schrödinger arrived at wave mechanics via de Broglie's matter waves, Heisenberg recognized that the difficulties concerning Bohr's theory were based in particular on the unscrupulous application of such ideas to atomic problems, which — as he realized — were impossible to test experimentally. Heisenberg therefore radically rejected to introduce terms and conceptions into his quantum mechanics of atoms which were not verifiable experimentally. For example, in the alternative theory of Schrödinger the introduction of potential energy is still necessary, and this is characterized by the idea of point-like nuclei and point-like electrons (Coulomb's law) ...* "[44].

Quantum phenomena are very different from those in classical physics. Therefore, it must appear as problematic to use classical concepts — here the classical concept of potential energy in connection with the potential function $U(x, y, z)$. Within the projection theory there cannot exist potential energies in $(\mathbf{r}, t)$-space because there are no real bodies embedded in $(\mathbf{r}, t)$-space. Furthermore, in contrast to the usual quantum theory, the classical function $U(x, y, z)$ gets a quantum-theoretical aspect within the projection theory, and this is because the system-specific time t has to be considered. $U(x, y, z)$ is automatically extended by the variable t and we have $V(r, t)$ (see in particular Sec. F.4). In other words, when we go from the classical theory to the quantum-theoretical description we have the transition $U(x, y, z) \rightarrow V(r, t)$ within the projection theory.

F.12.8 Summary

The notion "interaction in space" becomes problematic in conventional physics when we analyze the situation in terms of Newton's original conception but also within the usual quantum theory. According to Thomas Kuhn gravity has to be considered as an occult quality. The same problems come up when we consider the interaction between electrical charges as in the case of the hydrogen atom.

Within the projection theory the notion of "interaction" is a clear conception: In $(\mathbf{r}, t)$-space there are no interactions possible and defined, respectively. Here we have merely certain $\mathbf{r}, t$-correlations (geometrical structures with respect to the variables $\mathbf{r}$ and t) and there can be no $\mathbf{p}, E$-states in $(\mathbf{r}, t)$-space. No information or energy transfer through this space can take place. Within the projection theory the interaction between individual systems exclusively takes place in $(\mathbf{p}, E)$-space in the form of $\mathbf{p}, E$-fluctuations.

The $\mathbf{p}, E$-fluctuations and $\mathbf{r}, t$-correlations are the primary effects and not the interaction potentials which play at best the role of auxiliary elements for the description of $\mathbf{p}, E$-fluctuations and $\mathbf{r}, t$-correlations and, therefore, the interaction potentials have to be considered as secondary in character.

REFERENCES

1. Wolfram Schommers, QUANTUM PROCESSES, World Scientific, New Jersey, London, Singapore, 2011.
2. Wolfram Schommers, COSMIC SECRETS, Basic Features of Reality, World Scientific, New Jersey, London, Singapore, 2012.
3. C. G. Jung, Synchronizität, Akausalität und Okkultismus, Deutscher Taschenbuch Verlag GmbH & Co. KG, München, 1990.
4. John D. Barrow, The Artful Universe, Little, Brown and Company, Boston, New York, Toronto, London, 1995.
5. A. Einstein, Grundzüge der Relativitätstheorie, Verlag Vieweg, Berlin, Heidelberg, 1964.
6. O. Heckmann, Sterne, Kosmos, Weltmodelle, Deutscher Taschenbuch Verlag GmbH & Co. KG, München, 1980.
7. Brian Greene, The elegant Universe, Vintage Books, A Division of Random House Inc., New York, 1999.
8. K. C. Cole, The Hole in the Universe, A HARVEST BOOK, HARTCOURT Inc, San Diego, New York, London, 2001.
9. Nicholas Rescher, The Limits of Science, University of California Press, Berkeley, Los Angeles, London, 1977.
10. Gottfried Falk, Physik, Zahl und Realität, Birkhäuser Verlag, Basel, 1990.
11. W. Heisenberg, Manuscript zu einem Vortrag, den der Autor (Heisenberg) 1974 in Dubrovnik gehalten hat.
12. Philip J. Davies, Reuben Hersh, The Mathematical Experience, Birkhäuser Verlag, Basel, 1981.
13. Karl Popper/Franz Kreuzer, Offene Gesellschaft (Offenes Universum), Piper, München, 1986.
14. Siegfried Lenz, Über Phantasie, Hoffmann und Campe Verlag, Hamburg, 1982.
15. Wolfram Schommers, NANODESIGN. Some Basic Questions, World Scientific, New Jersey, London, Singapore, 2014.
16. Hoimar von Ditfurth, Der Geist fiel nicht vom Himmel, Deutscher Taschenbuch Verlag, München, 1980.

17. Paul Watzlawick, P.M. Perspektive Kommunikation, July 1989.
18. W. Schommers, Quantum Theory and Pictures of Reality, Springer-Verlag, Berlin, Heidelberg, New York, 1989.
19. Ivar Ekeland, Mathematics and the Unexpected, University of Chicago Press, Chicago and London, 1988.
20. Michael Talbot, Mysticism: The New Physics, Routledge & Kegan Paul, London and Henley, 1981.
21. Michael Talbot, The Holographic Universe, Grafton Books, London, 1991.
22. Wolfram Schommers, Advanced Science Letters 1, 59, 2008.
23. W. Schommers, Space and Time, Matter and Mind, World Scientific, New Jersey, London, Singapore, 1994.
24. W. Schneider, Hypothese, Experiment, Theorie, Sammlung Göschen, Walter de Gruyter, Berlin, New York, 1978.
25. Konrad Lorenz, Behind the Mirror, Methuen, London, 1977.
26. W. Schommers, The Visible and the Invisible, World Scientific, New Jersey, London, Singapore, 1998.
27. Jim Al-Khalili, Black Holes, Worm Holes and Time Machines, Institute of Physics, Bristol, 1999.
28. Hubert Goenner, Einstein's Relativitätstheorien, Verlag C. H. Beck OHG, München, 1997.
29. R. Brückner, Das schielende Kind, Schwabe Verlag, Basel, Stuttgart, 1977.
30. Thomas Kuhn, The Structure of Scientific Revolutions, University of Chicago Press, London, 1962.
31. K. Gödel, Rev. Mod. Phys. 21, 447, 1947.
32. Banesh Hoffman, Albert Einstein — Creator and Rebel, The Viking Press, New York, 1972.
33. Peter Graneau, Neal Graneau, In the Grip of the Distance Universe, World Scientific, New Jersey, London, Singapore, 2006.
34. A. Einstein, Grundzüge der Relativitätstheorie, Verlag Viehweg und Sohn, Braunschweig, 1973.
35. M. Born, Die Relativitätstheorie Einsteins, Springer-Verlag, Berlin, Heidelberg, 1964.
36. H. Dehnen, in Philosophie und Physik der Raum-Zeit, J. Audretsch und K. Mainzer (Hrgs.), Wissenschaftsverlag, Mannheim, Wien, Zürich, 1988.
37. B. Kanitscheider, Kosmologie, Philipp Reclam jun., Stuttgart, 1984.
38. M. Berry, Principles of Cosmology and Gravitation, Adam Hilger, Bristol and Philadelphia, 1989.
39. G. Falk, W. Ruppel, Mechanik, Relativtät, Gravitation, Springer-Verlag, Berlin, Heidelberg, 1983.
40. L. de Broglie, Die Elementarteilchen, Goverts, Hamburg,1943.
41. L.D. Landau, E.M. Lifschitz, Quantum Mechanics, Pergamon, Oxford, 1965.
42. M. Bunge, Canadian Journal of Physics 48, 1410, 1970.
43. F. A. Wolf, Taking the Quantum Leap, Harper and Row, San Francisco, 1981.
44. W. Finkelnburg, Einführung in die Atomphysik, Springer–Verlag, Berlin, Heidelberg, 1962.

Index

Printed in the United States
By Bookmasters